P9-ECN-413

ADVANCES IN SECURITY TECHNOLOGY

ADVANCES IN SECURITY TECHNOLOGY

Selected Papers of the Carnahan Conferences on Security Technology 1983–1985

Edited by

Romine (Dick) Deming, Ph.D.
Professor (Security Programs), College of Criminal Justice, Northeastern University, Boston, Massachusetts

Butterworths
Boston London Durban Singapore Sydney Toronto Wellington

Library of Congress Cataloging-in-Publication Data

Carnahan Conference on Security Technology.
Advances in security technology.

Bibliography: p.
Includes index.
1. Electronic security systems—Congresses.
I. Deming, Romine R. II. Title.
TH9737.C37 1987 621.389 87-9371
ISBN 0-409-90052-4

Butterworth Publishers
80 Montvale Avenue
Stoneham, MA 02180

10 9 8 7 6 5 4 3 2 1

Printed in the United States of America

CONTENTS

CONTRIBUTORS

David L. Andrew
R&D Manager, Arrowhead Enterprises, Inc., New Milford, Connecticut 06776

Rexford G. Booth
U.S. Army Belvoir Research and Development Center, Fort Belvoir, Virginia 22060

George K. Campbell
Analytical Systems Engineering Corporation, Burlington, Massachusetts 01803

Richard D. Capello
De La Rue Printrak, Inc., Anaheim, California 92807

Eddie Claiborne
Science and Engineering Associates, Inc., Albuquerque, New Mexico 87190

Douglas J. Clarke
Computing Devices Company, Ottawa, Ontario K1G 3M9

Ronald W. Clifton
Computing Devices Company, Ottawa, Ontario K1G 3M9

John Darby
Science and Engineering Associates, Inc., Albuquerque, New Mexico 87190

Romine (Dick) Deming
Professor (Security Programs), College of Criminal Justice, Northeastern University, Boston, Massachusetts 02115

H. Eugster
Cerberus Ltd., CH-8708 Männedorf, Switzerland

Hayden A. Flaugher
Program Manager, Plant Facilities and Systems Division, Argonne National Laboratory, Argonne, Illinois 60439

Harry D. Frankel
U.S. Immigration and Naturalization Service, Washington, D.C. 20536

Peter E. Green
Department of Electrical Engineering, Worcester Polytechnic Institute, Worcester, Massachusetts 01609

Kurt Gugolz
Cerberus Ltd., CH-8708 Männedorf, Switzerland

H. Güttinger
Cerberus Ltd., CH-8708 Männedorf, Switzerland

John R. Hall II
Analytical Systems Engineering Corporation, Burlington, Massachusetts 01803

R. Keith Harman
President, Senstar Corporation, Kanata, Ontario

Joseph R. Haumann
R.E. Timm and Associates, Inc., Hinsdale, Illinois 60521

R.E. Hilderbran
De La Rue Printrak, Inc., Anaheim, California 92807

Zol Kravets
Principal Electrical Engineer, Security Systems, Stone & Webster Engineering Corporation, Cherry Hill, New Jersey 08034

Edwin H. Morton, P. Eng.
Senstar Corporation, Ottawa, Canada

G. Pfister
Cerberus Ltd., CH-8708 Männedorf, Switzerland

Francis P. Pfleckl
Project Engineer, U.S. Army MERADCOM, Counter Surveillance/ Counter Intrusion Laboratory

Jay A. Rarick, Sc.M.
U.S. Army Belvoir Research and Development Center, Combined Arms Support Laboratory, Physical Security Equipment Division, Fort Belvoir, Virginia 22060

Donald L. Reigle
Plant Facilities and Systems Division, Argonne National Laboratory, Argonne, Illinois 60439

Stephen Riter
The University of Texas at El Paso, El Paso, Texas 79968

Ronald R. Rudolph
R.E. Timm and Associates, Inc., Hinsdale, Illinois 60521

Michael S. Sims
Computing Devices Company, Ottawa, Ontario K1G 3M9

Peter Smith
Science and Engineering Associates, Inc., Albuquerque, New Mexico 87190

P. Steiner
Cerberus Ltd., CH-8708 Männedorf, Switzerland

Willard Thomas
Science and Engineering Associates, Inc., Albuquerque, New Mexico 87190

Ronald E. Timm
R.E. Timm and Associates, Inc., Hinsdale, Illinois 60521

P. Wägli
Cerberus Ltd., CH-8708 Männedorf, Switzerland

Scott Walker
Science and Engineering Associates, Inc., Albuquerque, New Mexico 87190

Thomas E. Zinneman
R.E. Timm and Associates, Inc., Hinsdale, Illinois 60521

PREFACE

As a professor of security technology and a long-term participant in Carnahan Conferences on Security Technology, I realize the direct value of many of the papers to security administrators and students. I am aware of a recent poll of security administrators indicating the desire to know more about security technology ranked second in the area of information desired. Information on security law ranked first. The small but faithful group of security administrators who attend the conferences, as friends and confidants, have also inspired this project. I have become convinced of the value to security administrators of collecting the more directly relevant and understandable papers into a book where they would be easily accessable to other administrators.

The valid objective of the Carnahan Conferences is to share theoretical positions and report on research, development, and design results in security technology. This naturally attracts participants with strong backgrounds in physics, engineering, mathematics and other physical sciences. Most of the articles are abstract, mathematical and/or very specific. These backgrounds are not typically those of security administrators. Their backgrounds tend to be in administration, management or the behavioral sciences. Although engineering and science is necessary for design of devices and technical systems, one does not have to be an engineer to be a knowledgeable user of security technology. The application of products is a different matter.

Many of the articles are more qualitative by virtue of the focus of the article or the maturity of the product or system. Some of these articles are more focused on application, with security administrators as the audience addressed. In short, most of the articles appropriately address what is in the ''black box'' or how to relate black boxes together into systems of security technology. Yet, some articles address the application to problems which are of interest to security administrators. It is these articles I have tried to select for this compendium. Those wishing exposure to a greater range of articles are encouraged to write for copies of the proceedings. Copies are available for all the conferences since 1968, and may be purchased by writing to:

OES Publications
226 Anderson Hall
University of Kentucky
Lexington, Kentucky 40506-0046

My participation in the Carnahan Conferences on security has greatly benefitted me as a professor in security technology. I hope that my participation has benefitted my students. I also must confide that my association has been a personal source of enjoyment. The chance to exchange ideas formally and informally in an outstandingly hospitable environment is an annual inspiration.

My gratitude first must be expressed to all those with whom I have been associated during these years. Space does not permit me to recognize and thank them individually. However, some people must be singled out for establishing a most conducive environment for our interchange. Dean David Blythe, Professor John Jackson, Cheryl Banks, and Ili Hall are sincere participants in the art of "Southern hospitality." Pieter de Bruyne, chairman, and the International Organizing Committee are commended for promoting the outstanding conference in Zurich in 1983. The twenty-two members of the Executive Committee must be complimented for putting together superb conferences and also for facilitating a productive and enjoyable conference environment.

These articles would not be available in any form without the efforts of Dean David Blythe and John Jackson, the proceedings editors, and William DeLore, the proceedings publishing editor. Special thanks are here conveyed to Greg Franklin, editor at Butterworths, not only for his invaluable counsel on this project, but for his help on other projects over the past ten years. Of course, the primary gratitude for this project goes to the authors of the papers included in this compendium. My only regret is that I could not include more.

INTRODUCTION

''What's in a name?'' The Carnahan Conferences on Security Technology have been synonymous with state-of-the-art security technology. It is where the interchange of knowledge occurs through the presentation of learned papers. It is where engineers, physicists and other scientists have shared the results of their research and product development with each other on countermeasures of crime and promotion of security for the past 20 years. Security administrators are, immediately or eventually, the beneficiaries.

The conference derives its name from the facility in which the annual conferences are held. The Carnahan House, the University of Kentucky's conference center, was purchased by the university in 1956 along with the remainder of the estate long known as Coldstream Farm. The property is situated in the heart of the bluegrass country just north of Lexington, the geographical center of the state. Coldstream Farm has a prestigious history associated with swift horses.

The farm, when owned by Price McGrath, produced the first Kentucky Derby winner, Aristides, who became immortal with his 1875 presence in the winner's circle. Colonel Milton Young bought the farm after the death of Mr. McGrath and continued the tradition. The farm became the home of Hanover, who sired many of the best horses of the day. His blood stills flows through the veins of many contemporary winners.

The heritage was continued by subsequent owners. During the farm's heyday, it was the center of social life of horse people. It was a place of excitement and comradeship, where lavish parties were given and sumptuous barbecues were enjoyed. It was where the best Kentucky bourbon flowed in association with sophisticated conversation on the ''sport of kings.'' In this tradition, conference participants still enjoy comradeship and barbecues while engaging in sophisticated dialogue of a different nature, security technology.

Coldstream Farm continues in the tradition of enhancing animal husbandry as a University of Kentucky agricultural experimental station. The Carnahan House was first used by the university as the Alumni-Staff Club. It was given its name in honor of an alumnus and generous benefactor James W. Carnahan, of the Lyons and Carnahan Book Company. In 1962 the club moved to a more suitable estate which had been recently acquired by the university, thus permitting the allocation of Carnahan House as an ideal conference center.

The Carnahan Conference on Security Technology is the oldest continuous conference exclusively devoted to security technology. In addition, it has documented the progress in security technology through its proceedings published since the 1968 conference. Copies of all the proceedings are still available.

The first conference was initiated by Professor Robert Cosgriff, chairman of the Department of Electrical Engineering at the University of Kentucky. President Johnson had just launched his "war on crime" and Professor Cosgriff felt that a conference on electronic countermeasures to crime would be a worthy contribution to the war. A regional conference was convened in 1967, drawing about a dozen people from Cincinnati, Ohio, Louisville and Lexington.

Three men took the initiative to organize a national conference for the following year: Professor John Jackson, who has also been the proceedings' editor from the beginning; Lawrence D. Sanson, president of the Lexington Chapter of the Institute of Electrical and Electronics Engineers (IEEE) and an employee of GTE; and Tandy Haggard who was employed by IBM. The conference continues to be sponsored by the College of Engineering, the Lexington Chapter of IEEE, and the Aerospace and Electronic Systems Society section of IEEE.

The first national conference became, in fact, an international conference by virtue of attracting participants from other countries. The annual meetings continue to attract participants from as far away as Japan, Taiwan, Switzerland, Sweden, and West Germany. Many participants come from Canada and the United Kingdom. In addition to the annual conferences, international conferences are held every third year in another country. Five international conferences will have been held by 1986: 1973 Edinburgh, Scotland; 1977 Oxford, England; 1980 West Berlin, Germany; 1983 Zurich, Switzerland; and 1986 Göthenburg, Sweden.

Currently there are many conferences and workshops devoted to security technology. Yet, the Carnahan Conferences on Security Technology continue to bring together those involved in security technology research and development, users, and professors for an interchange of ideas on the cutting edge of securing people and property from those who would steal, maim, kill, and destroy.

ADVANCES IN SECURITY TECHNOLOGY

PART I

Security Planning

The days of simple organizations are fast fading. Today's grocery stores are more likely to be complex organizations than were grocery stores of the past. Certainly manufacturing companies, hospitals, and universities, as well as many other organizations, have become highly complex. Their security problems are similarly complex. No longer is the burglar the essential threat and the money safe the sufficient countermeasure.

Now the list of threats is much greater, encompassing internal thieves as well as external thieves. Today, espionage agents steal bits of information or prototypes to pass to enemy countries or competing firms. Saboteurs will damage production capabilities and vandals will vent frustration by damaging delicate machinery or electronic data processing capabilities. Terrorists are determined to harm targeted organizations and people. The threat of a civil suit, because security needs were neglected, looms constantly in the background.

Accidents are more prone to occur. Hazardous material may accidentally spill injuring personnel. Accidents in data processing may destroy computer programs or stored data. Even the range in natural disasters has expanded and the impact of all disasters is more devastating. Companies can be paralyzed during times of power failure. And the list goes on, seemingly to infinity.

The range in security solutions has likewise become more complex to meet the expanding range of problems. The components of comprehensive security systems: policy, procedure, personnel, and products have necessarily become more complex. No longer sufficient is a policy for the manager to lock the safe. Now the security policies for many organizations are voluminous.

Procedures have also multiplied at a geometric rate. No longer will the directions for locking the safe and the combination for unlocking suffice. Numerous security procedures exist for all staff, but especially for security personnel. The procedural manuals for even relatively small organizations are the size of a large city telephone directory.

The demand has increased for specialized personnel to implement policy and execute procedures. The number of people and the level of education is necessarily much higher. At least ten universities offer masters programs in security administration, at least twenty-five offer bachelors programs and eighty-five offer associate degrees.

The number and complexity of security products has mushroomed in the last ten to twenty years. The demand by consumers for highly technical devices and the concomitant response by producers has generated a huge increase in the technical knowledge available for the design of highly technical and complex security devices and systems. This magnitude of devices and options inspired the Carnahan Conferences in Security Technology, and the complexity of options is reflected in the pages of this book.

In short, contemporary organizations have become exceedingly complex. The attendant security problems and solutions are likewise exceedingly complex. How does the security administrator/planner analyze these complex security problems and plan security solutions successfully? Nothing can be missed. It is the accidently omitted vulnerability which will be discovered and compromised. The complex elements of security components must be precisely weaved together to capture synergy. Synergy, when the sum of its parts is greater than the whole, is a resource to those who have the ability to precisely weave intricate elements together to form systems to achieve a given objective.

Appropriately, this first section contains two articles addressing the problem of analyzing complex security needs and planning comprehensive security systems. The two teams of authors appropriately address the fact that analysis must be complete and economical. They stress also that security systems must not only be the most beneficial available solutions, but they must be cost effective as well.

The two teams of authors have gained valuable knowledge in the trenches, so to speak. By being members of security planning firms, they have been required to analyze security problems and design systems for complex organizations serving national needs. Their experiences have generated effective strategies relevant to any more-or-less complex organization. We are the beneficiaries of these strategies.

Technical Methods to Optimize Protection against Sabotage and Theft

John Darby
Willard Thomas
Eddie Claiborne
Peter Smith
Scott Walker

Abstract. This article describes how to optimize security system design for a complex facility or operation using sophisticated techniques originally developed for nuclear-related security. The following topics are discussed: identification of what to protect; evaluation of security system options; selection of security system components.

ACKNOWLEDGMENTS

The authors of this report wish to acknowledge Richard Worrell of Sandia National Laboratories (SNL) who developed the SETS computer code, and Miller Cravens, Al Winblad, and Carl Clark, also of SNL, who developed the adversary sequence diagram concept and the PANL computer code.

INTRODUCTION

For the past few years we have been heavily involved in nuclear safeguards and security programs sponsored by the United States Government. We have performed sophisticated safeguards/security analyses for over 40 US nuclear facilities. In an earlier paper,[1]

1983 International Carnahan Conference on Security Technology, Zurich, Switzerland October 4–6, 1983.

we described how sophisticated techniques used to optimize security at nuclear installations can be directly applied to any complex facility, process, or information handling network. The intent of this paper is to discuss the needs for such techniques, and to provide a brief review of these techniques.

Four major steps are required to ensure the proper design and operation of a security system for a complex facility or operation. These are listed below:

1. Identification of entities requiring protection.
2. Examination of security system options and selection of the optimum system.
3. Selection of specific components for the optimum security system.
4. Installation and maintenance of the security system.

Steps 1 through 3 will be discussed in this paper. Also, a simple application of steps 1 and 2 is presented.

ANALYSIS PROCEDURE

Identification of What to Protect

For a complex facility or network, it is not trivial to determine a minimum yet complete set of entities to protect. (Entities are items, materials, information, and so on.) Completeness can be assured by protecting everything, but this is neither cost effective nor necessary. To minimize the number of entities to protect, some sort of analysis must be performed.

For a simple facility, this analysis can be a trivial process. A small jewelry store obviously identifies valuable jewels as entities to protect. For complex facilities, a more rigorous analysis is necessary. A large bank concerned with protection of assets must not only protect money physically on site, it must also protect against unauthorized electronic transfer of funds through its interactive computer system.

One approach to determine what to protect at a complex facility is to answer the following two questions: What undesirable results am I concerned about? and What entities need to be protected to prevent each undesirable result? Undesirable results are those identified to be of serious concern to the facility owner such as "Theft of Proprietary Information" and "Sabotage of Production Capability."

It is frequently possible to directly apply a deductive analysis to determine the entities to protect. (A deductive method determines *how* a given undesirable result can occur by identifying "failures" that lead to the result.[2]) For example, "Loss of Power to Computer" is an undesirable result that can be analyzed by Fault Tree Analysis, a widely used deductive technique. Computer codes can assist in the analysis to identify which sets of entities may be protected to preclude the undesirable result. (For example, see reference 3.)

It is possible that an undesirable result is too difficult to analyze directly by a deductive approach. In this case, the result should be more explicitly defined. If a deductive approach is still intractable, an inductive approach can be used. (An inductive method determines *what* results arise from given "failures."[2]) For example, "Economic Loss" is an ill-defined result. Suppose that "Economic Loss" can be more explicitly described by "Loss of Production for Greater than a Critical Time." If the facility is very complex using many operations to produce a variety of products, even this more succinctly defined result is impractical to analyze deductively down to an entity (component) level. An inductive analysis can be used to identify critical portions of the facility. For example, one type of inductive analysis is to assume loss of any Input (raw materials, utilities), Operation (step performed within the facility), or Output (storage of critical products) and determine which losses result in downtime exceeding a critical value. (This is a Failure Mode and Effects Analysis.[2]) To determine whether or not a given loss results in excessive downtime, seven types of times can be defined (time to restore an operation if it is lost, time to replace an input storage to minimum operating level, and so on); then, a simple computer program can be used to identify critical losses.[4]

Once the "system" level losses of concern have been identified, a deductive technique can be used to determine specific combinations of component level "failures" that result in "system" loss.

In these security-related analyses of what to protect probabilities are not considered. This is because the probability that a security-related "fault" (sabotage, theft, or other malicious act) is successful is dependent on willful intent, and on the security system in place to prevent the "fault." Another technique, subsequently described, examines how well the vulnerabilities can be protected. [Because all events in effect have a probability of "1," common mode effects, so crucial to traditional probabilistic analyses, are not of major concern.[5]]

Evaluation of Security System Options

After an appropriate set of entities to protect has been identified, security system options for protecting these entities should be proposed and compared. Computer codes can be used for this comparison.[6] To quantify the comparison, threats must be specified and protection elements for the proposed security systems identified.

Physical security elements consist of: walls; doors and windows; intrusion detection devices; physical barriers; alarm assessment devices; contraband detection devices; personnel detection devices; and control and communications devices.[7,8,9,10] These elements combine in various ways to counter the following types of threats: outsiders with various attributes (weapons, explosives, vehicles, etc.); outsiders in collusion with insiders (facility employees); and insiders with or without contraband (weapons and/or explosives).

Information security elements consist of: passwords, logs, audits, encryption/decryption devices, monitoring software, procedures, and selected physical security elements.[11,12] These elements combine in various ways to counter the following types of threats: outsiders; insiders without access to files of concern; insiders with access to files of concern; and outsiders in collusion with insiders.

One measure of the effectiveness of a given security system design under a given threat is the probability of detecting adversary actions while there is yet time for effective response. This measure, "timely detection," allows for cost-effective comparison among many security system options under a wide range of threats (overt terrorist sabotage, covert employee theft, and so on).

Alternative security designs are investigated by simply adding or removing security elements to or from the system. This process has two advantages:

1. The ability to alter the effectiveness of the security system, and secondly,
2. the ability to control some dependent variable of choice such as cost or operational impact.

Judicious manipulation of these two factors allows the analyst to optimize a security system based on whatever criteria are deemed important. The results of the analysis can provide a measure of the cost in dollars to achieve an extra increment of security or can simply be used to design a system of equivalent effectiveness at lower cost.

It should be emphasized at this point that it is not necessary to obtain an absolute measure of security system performance. The results of the analysis are important from the standpoint of a comparison among two or more design options. Due to the inherent uncertainties involved in quantifying adversary actions, differences in system effectiveness are considered significant only if they are greater than an order of magnitude. With this in mind, it can be seen that the analysis has two principal applications:

1. To verify that no paths are "free," that is, not protected.* (This is a nontrivial application for complex facilities.)
2. To identify those hardware/software elements that provide a significant enhancement in security at the specific facility.

Large holes in a security system become evident during the analysis where before they may have gone unnoticed. Likewise, weaknesses and strengths tend to become obvious during the analysis. This has the added benefit of indicating where inefficiencies have arisen due to overprotection.

Based on the analysis results, elements of a selected security system are identified in terms of performance requirements (time delays and detection probabilities). Security system design or upgrade then becomes an iterative process of proposing hardware/software changes and evaluating proposed enhancements. For each security system option that is considered, cost, operational impact, safety, and performance are evaluated to select the optimum system to be installed.

---

*Paths are sequences of adversary actions directed at protected entities.

Selection of Security System Components

As a result of the previously described analyses, security system elements which protect specific entities have been identified. Also, performance requirements for these elements have been produced. An element is a portion of a security system which fulfills a specific function. For example, "intrusion detection" is an element to provide the function "detect unauthorized penetration." A component is a specific hardware or software item. For example, a "hand geometry reader" may be the component selected to provide the performance necessary for the element "personnel identification;" or, a "doppler microwave detector" component may satisfy the element "interior intrusion detection."

Given that an element has been specified for a security system, the specific component to satisfy the required performance must be selected. This selection is based upon established characteristics of candidate components coupled with facility-specific constraints, such as: environment, operational interface, and safety. In order to properly select specific components, the security engineer must ensure that the components perform satisfactorily in their intended application. Although the literature can assist in proper selection (see references 7 through 13), it is mandatory that the security engineer have an extensive knowledge of state-of-the-art security hardware/software, plus extensive experience in procuring, installing, and maintaining complex security systems.

A Simple Example

A medium size electronics firm is heavily involved in contract bidding with a number of strong competitors. To bid as low as possible and yet maintain desired profits, the company maintains accurate records of its volume of business, projected work, and its cash flow position. This information is called "profit record information" by the company, and it is crucial to competitive performance. The profit record information is recorded in three places at the company sales office:

1. A hard disk connected with the interactive computing system at the company contains almost real-time information used to provide guidance for engineer/sales staff bid proposals. One password, known to many employees, allows remote access to all disk files.
2. A backup magnetic tape is provided in case the disk is damaged. This tape is updated every three months.
3. A set of master bookkeeping records is kept in the company safe for auditing purposes. This record contains manual entries and is updated monthly as end-of-the-month account balances become available.

The company has decided to analyze the security of its profit record information and to consider the cost-effectiveness of modest upgrades. At present, security is

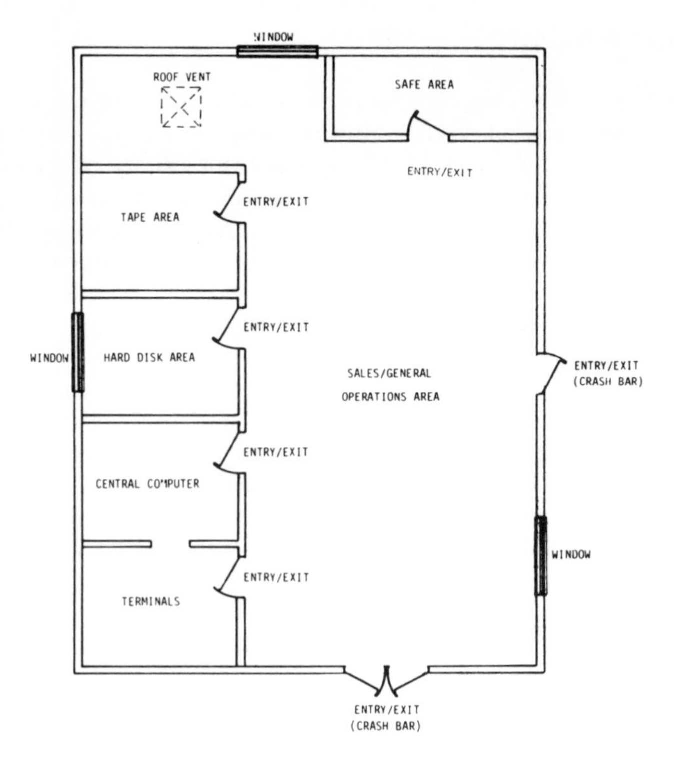

Figure 1. Facility layout.

provided by key locks and one on-site contract guard. Figure 1 illustrates the layout of the sales office.

Identification of What to Protect

Based on the information presented in the preceding example, one can construct the sabotage fault tree shown in Figure 2. Each sabotage event is represented by a rectangle, along with a brief prose description. Other symbols used in the tree are shown below:

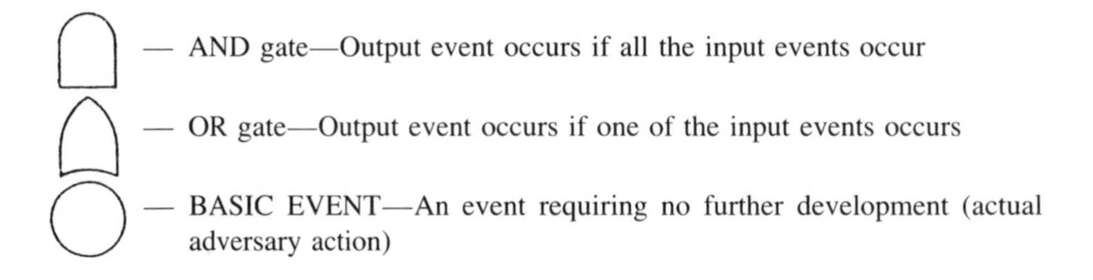

— AND gate—Output event occurs if all the input events occur

— OR gate—Output event occurs if one of the input events occurs

— BASIC EVENT—An event requiring no further development (actual adversary action)

The event solution for the sabotage tree contains three separate combinations of adversary event actions. These are:

1. A and B and C, or
2. A and B and D and E, or
3. A and B and D and F.

To protect against successful sabotage, the protection of any one of the following events or event combinations will be sufficient:

1. A, or
2. B, or
3. C and D, or
4. C and E and F.

It is often useful to assign locations to each basic event as shown in Figure 2. This allows a determination of the location an adversary must visit to accomplish his goal. Note that events D and E have been assigned to areas "OUTSIDE." This indicates that these actions can be accomplished from outside the facility. The "location solution" of the tree is as follows:

1. SAFE and TAPE and DISK, or
2. SAFE and TAPE and OUTSIDE, or
3. SAFE and TAPE and OUTSIDE and TERMINAL.

Since one can only protect areas within the facility, the solution reduces to

1. SAFE and TAPE and DISK, or
2. SAFE and TAPE, or
3. SAFE and TAPE and TERMINAL.

Furthermore, items 1 and 3 now disappear since they involve additional (and thus unnecessary) adversary visits to DISK or TERMINAL. Therefore, the final location solution is simply

SAFE and TAPE.

To ensure destruction of the information the adversary must visit both SAFE *and* TAPE. Therefore, to prevent sabotage, the facility must protect either the SAFE *or* TAPE location. Protection of the DISK and/or TERMINAL areas does not serve any function, since the disk information can be destroyed from external locations which are impossible to protect. (The facility password is assumed to be easily compromised.)

For the simple example just presented, it was possible to obtain complete event and location solutions by inspection. Fault trees representing actual facilities typically contain several hundred logic gates and basic events. Some solutions to the large trees

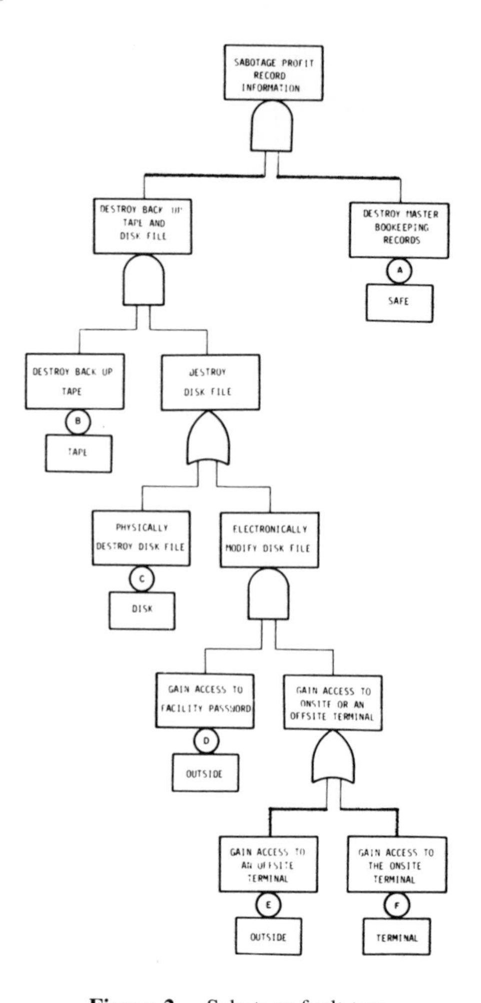

Figure 2. Sabotage fault tree.

can be generated by hand. However, a computer analysis is required to obtain a complete set of solutions. In addition, the computer can be used to optimize security protection requirements from, say, an economic standpoint. The SETS computer code has been used to perform analyses of complex systems.[3]

A theft fault tree representing our simple example is shown in Figure 3. The location solution is given by the following:

1. SAFE, or
2. TAPE, or
3. DISK, or
4. OUTSIDE and TERMINAL, or
5. OUTSIDE

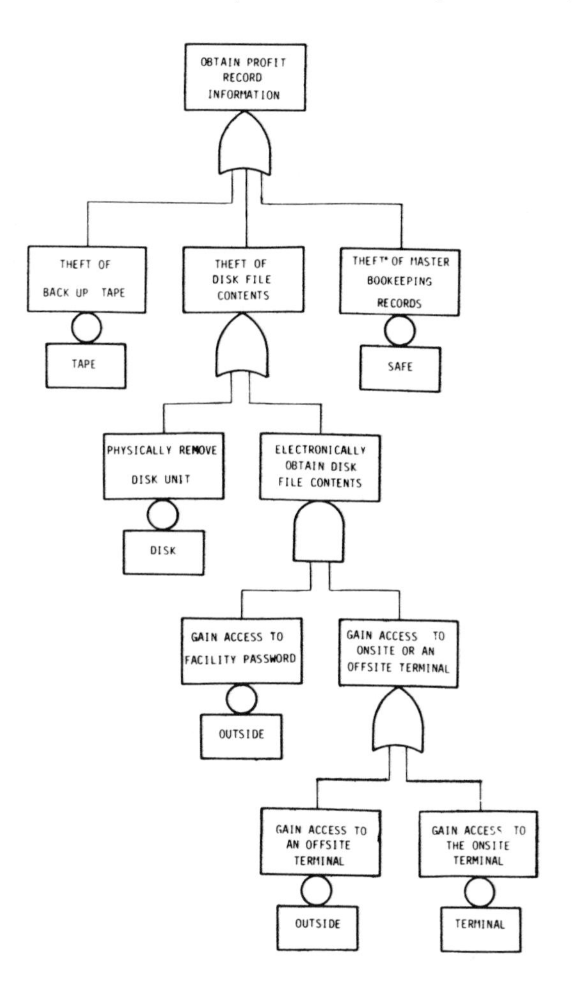

Figure 3. Theft fault tree.

Since the theft of disk information can be accomplished from outside, protection of facility areas is not adequate. This is an entirely different situation from that encountered in the case involving sabotage. (Theft can be accomplished from outside the facility because the facility password is assumed to be easily compromised.)

Evaluation of Security System Options

To demonstrate a simple method for evaluating a Security System, the Adversary Sequence Diagram (ASD) approach, originally developed at Sandia National Laboratories,[6] will be applied to the company profit record example.

Assume that the company has opted to prevent sabotage by protecting *one* copy of the profit record, the magnetic tape. To prevent theft of the proprietary information, *all* copies of the profit record must be protected—the tape, the work disk, and the master bookkeeping record in the safe. In this example, more areas need to be protected against theft than sabotage; however, this does not automatically mean that theft protection considerations lead to a higher level of protection than do sabotage considerations. Note that to prevent sabotage, the adversary must be kept *out* of the tape area while access to the tape, disk, or safe does not in itself constitute theft; the adversary must "take" the information from the facility to steal it.

The ASD represents facility areas by rectangles and paths joining areas by lines. For the facility in question, the ASD's for sabotage and theft are shown in Figures 4 and 5, respectively. The ASD for theft is more complex than the one for sabotage for two reasons: 1) there are more theft targets than sabotage targets, and 2) the ASD covers entry into theft targets *and* exit from theft targets. Note that theft can be accomplished by two different methods: "hands on" theft from the tape, disk, or safe areas, or reading information from the disk file given appropriate access (passwords). (It is assumed that the computer system is interactive and can be accessed by terminals both onsite and offsite.)

The ASD can be used to ensure that a given threat of concern does not have "easy" access to items to be protected. The elements of a given security system are associated with appropriate lines on the ASD. (Those elements shown in Figures 4 and 5 represent the existing security system at the facility.)

The Path Analysis (PANL) computer code can be used in conjunction with the ASD to identify critical paths offering the least resistance to specific threats for a

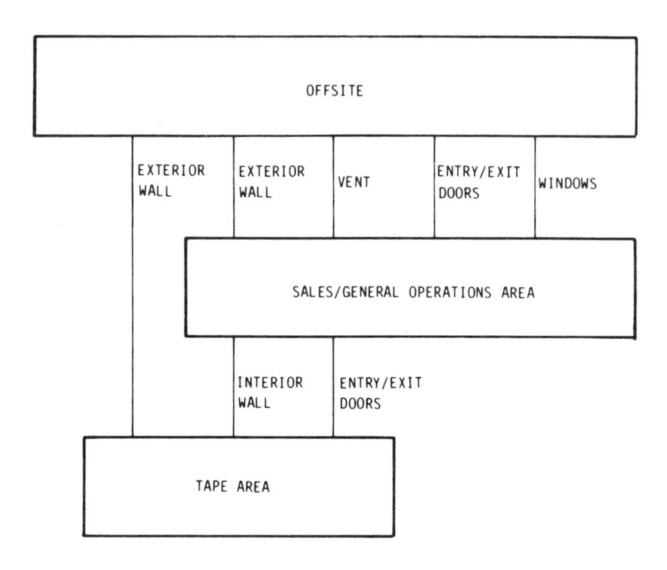

Figure 4. Adversary sequence diagram for sabotage.

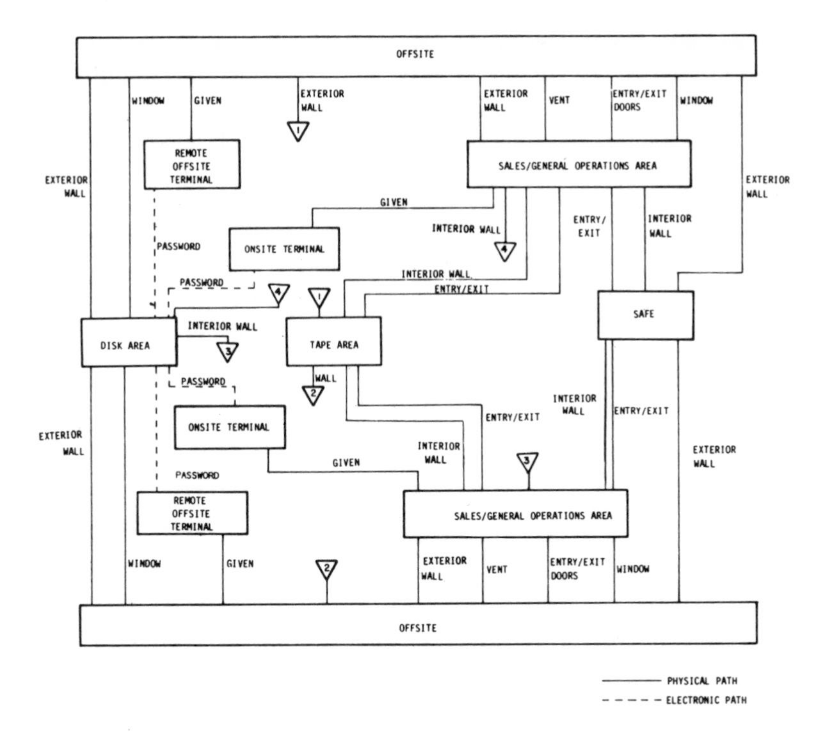

Figure 5. Adversary sequence diagram for theft.

specific security system.[6] For example, analysis of the ASD would indicate that the critical path for a non-employee (outsider) threat is:

Remote Offsite Terminal→Disk→Remote Offsite Terminal, because the password is not secure. Better control over electronic access to the sensitive disk file is necessary. By modifying security system elements along the lines, two types of security system characteristics can be quantified:

1. Ensure that for each threat of concern there is no *free* path to items that are supposedly protected.
2. Upgrade the *resistance* of paths to threats of concern.

For Step 2, a threat-specific data base is necessary. To be accurate, such a data base should be founded on actual threat-specific tests of security system components. For physical security system elements, an experienced security engineer can generate a data base using available information on elements' performances.

Using such a data base, PANL will quantify critical paths. The time delays and detection probabilities afforded by the elements of a given security system are associated with path segments between areas. Using the principles of conservatism, it is assumed that the adversary will minimize his chances of being detected until he is within one guard-response time of his target. With any chance of effective response thereby negated, the adversary will attempt to minimize any time delays associated with achieving his goal. Therefore, for a given security system design under a given threat, PANL computes paths with the lowest probability of detection while there is yet time for effective response. Note that the type of threat causes a fundamental change in the breaking point between detection and delay. The thief is interested in getting within one guard-response time of escaping from the facility while the saboteur is interested in just reaching his target. In mathematical terms, PANL computes

$$\prod_{i=1}^{K} \overline{P}_{i \text{ maximum}}$$

for

$$\sum_{i=1}^{K} T_i \geqslant T_R$$

and

$$\sum_{i=k+1}^{\text{end}} T_i < T_R.$$

The "i" denotes a path segment (a line on the ASD), $\overline{P}_i$ is the probability of non-detection for a given threat at the ith segment, T_i is the time delay encountered by the threat along the segment, and T_R is the response time (guards, audits, etc.).

CONCLUSIONS

These techniques have been applied to physical security of nuclear facilities. They are directly applicable to optimization of physical security for other capital-intensive, complex facilities such as: oil refineries, power plants, semiconductor device production plants, and government defense facilities.

These techniques can be used for information and data security as well. The issue of data security is generally broader than that of physical security. In fact, data security often requires physical security considerations. The identification of what to protect is more difficult since information and databases usually require a heirarchy of access with different types of privileges accruing to each level. But, if the security elements of a given system can be quantified, then the techniques can be applied with privileged domains taking the place of areas and electronic paths replacing physical ones. Modern data handling systems are in fact so complex that they cannot be secured without applying some kind of comprehensive methodology.

The dual nature of information security, physical and systemic, becomes most evident when the concern is sabotage. The saboteur can destroy information by electronic paths (erasure, rewriting, etc.) or he can physically destroy the hardware containing that information. Such considerations are generally important to companies concerned with industrial sabotage in the form of destruction of valuable information. Theft or compromise of information (especially illicit alteration of data) is usually more independent of physical considerations. Information dealers and institutions dealing with electronic transfers of funds or sensitive information are the principal targets of compromise and theft attacks. Of course, access to the central computer system of such an institution also constitutes compromise, so physical security is important, but the overriding concern is the security of sensitive domains.

It is evident, therefore, that data security for a complex system requires a methodology, such as that previously described, which can successfully integrate physical and systemic concerns.

RECOMMENDATIONS

It is recommended that sophisticated techniques be applied to optimize security at complex facilities or operations. If appropriately applied, the techniques described in this paper can ensure:

- complete identification of entities to protect;
- evaluation and selection of an optimum security system for a specific facility;
- selection of hardware/software components appropriate for facility-specific application.

The high costs associated with failure to thwart adversary threats make the cost of performing such analyses small indeed.

REFERENCES

1. Darby et al., ''Vulnerability Analysis and Integrated Security System Design: The Technical Approach,'' 1983 Conference on Crime Countermeasures and Security, University of Kentucky, Lexington, Kentucky, May 11–13, 1983.
2. Vesely et al., *Fault Tree Handbook,* NUREG-0492, January 1981.
3. Worrell, *A SETS User's Manual for the Fault Tree Analyst,* NUREG/CR-0465, 1978.
4. Memorandum of Record from J. L. Darby to D. L. Mangan, 25 March 1983.
5. *PRA Procedures Guide,* NUREG/CR-2300, January 1981.
6. Cravens et al., *Path Analysis (PANL) User's Guide,* SAND80-1888, September 1980.
7. *Intrusion Detection Systems Handbook,* Sandia Laboratories, SAND76-0554, October 1977.
8. *Entry-Control Systems Handbook,* Sandia Laboratories, SAND77-1033, September 1977.

9. *Barrier Technology Handbook,* Sandia Laboratories, SAND77-0777, April 1978.
10. *Safeguards Control and Communications System Handbook,* Sandia Laboratories, SAND78-1785, May 1979.
11. Hoffman, *Modern Methods for Computer Security and Privacy* (New Jersey: Prentice-Hall Inc., 1977).
12. Parker, *Computer Security Management* (Virginia: Reston Publishing Co., 1981).
13. Barnard, *Intrusion Detection Systems* (Boston: Butterworth Publishers, 1981).

Integrated Security System Definition

George K. Campbell
John R. Hall II

Abstract: The objectives of an integrated security system are to detect intruders with a high degree of reliability and then to deter and delay them until an effective response can be accomplished. Definition of an effective integrated security system requires proper application of classical system engineering methodology. This paper summarizes a classical system engineering methodology and describes the application of this methodology to the problem of integrated security system definition. The paper concludes with the product of the process: implementation of an integrated security system.

INTRODUCTION

The past approach toward the improvement of security programs at high-security locations has been one dictated by fluctuating budgets, intense competition for scarce resources, and periodic emphases to rapidly upgrade safeguards capabilities in response to identified vulnerabilities. This approach has largely mitigated against systematic overall considerations of cost benefit and effectiveness trade-offs for both required and proposed security upgrades. Improvements in one subsystem have been initiated at considerable cost without full knowledge of the contribution of that element to vulnerability reduction and safeguard readiness. If improvements were scrutinized individually as they were proposed and implemented, they might appear to have value in and of themselves. But, in the total context of an integrated approach to the effective reduction of vulnerability, little real security mission upgrade may be realized by these periodic infusions of new resources.

Our experience in the field acting in the roles of troubleshooters of already installed "integrated systems," leads us to conclude that system engineering methodology application to security system design is not well understood. More basically, it appears

1983 International Carnahan Conference on Security Technology. Zurich, Switzerland, October 4–6, 1983.

that the classical system engineering process, while often cited, is misunderstood and/or misapplied in the interest of the use of hardware in search of an application. Proper application of such a process is particularly critical to integrated security system design because of the importance of the synergism between subsystems and subsystem elements and the relative mix required for each site. In a security system, the effectiveness of the whole must be greater than the sum of the parts or new vulnerabilities will be created by the system "solution."

There is a need to define a process whereby security vulnerabilities can be analyzed, and the costs of systemic improvements can be calculated to reduce such vulnerabilities. The process should employ evaluatory tools which can aid both systems planners and resource allocation decision makers. It should have the objective of defining the contribution of new subsystem elements toward total system capabilities and effectiveness.

The purpose of this paper is to identify critical concerns in the development of a process for analysis and design of complex (multi-subsystem) security systems and to set forth a suggested approach.

METHODOLOGY OVERVIEW

Integrated security system definition necessitates a comprehensive systems approach due to the synergistic relationship among the various subsystems which comprise a fully integrated safeguard system.

The ultimate objectives of the system engineering process are to establish, optimize and implement the requirements for facilities, equipment, personnel and procedures. It is a systematic approach to achieve optimum overall system design and performance through accurate requirements definition, appropriate allocation of these requirements, effective functional definition of interoperable subsystems and procedures, cost-effective system design through selection of technically sound equipment and appropriate acquisition, leading to installation, test and full system implementation.

Initial focus must be upon the threat(s) confronting the assets to be protected. The security system designer must then take into account each respective safeguards subsystem which may be established to detect, deter, delay and respond to these threats. These macro subsystems can be initially identified as:

- the physical resources to deter and delay an adversary;
- equipment installed to detect and assess intrusion attempts and unauthorized activities;
- personnel utilized for security systems operations, management, and support;
- procedures essential for system operation and response effectiveness; and
- personnel equipment for security force support.

While these subsystems need to be individually evaluated, the synergistic interactions among them must be inter-related at a system level. It must be recognized that the integration of these safeguards resources does not involve a static set of environments. Because threat is dynamic, the fully integrated safeguards system must provide redundancy, diversity and flexibility in order to provide defense-in-depth.

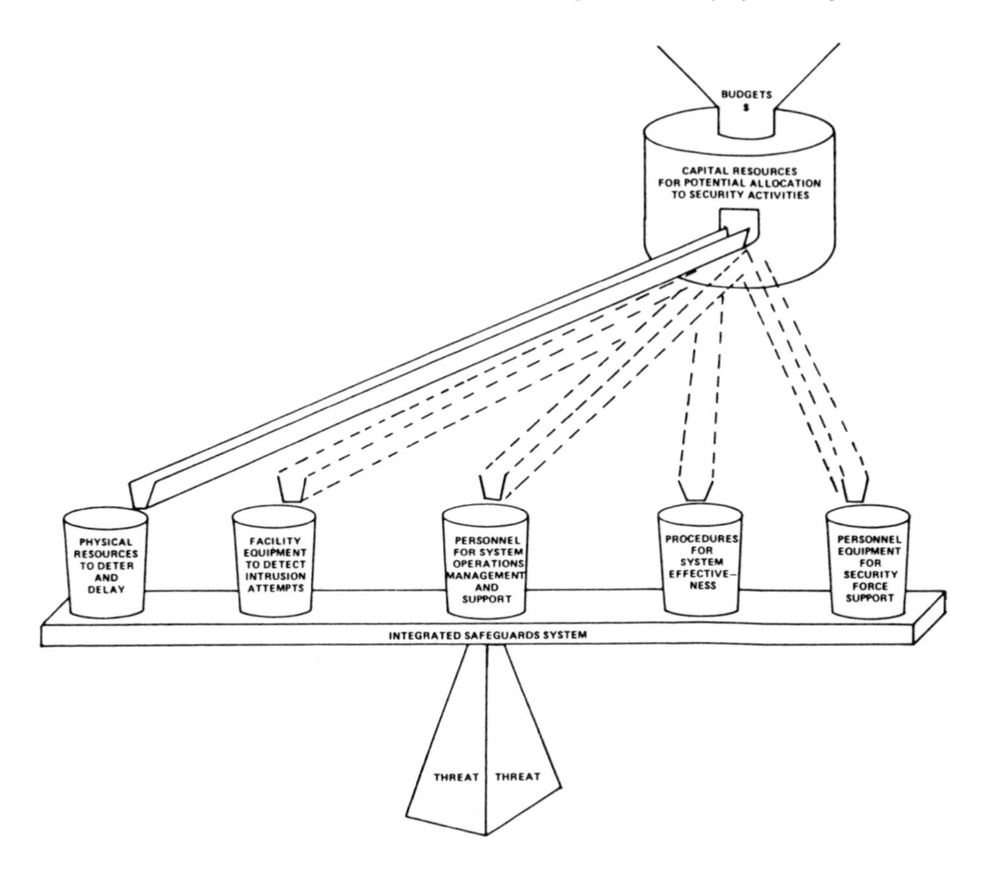

Figure 1. Factors determining the integrated security system and their relationship with one another.

Finally, a capital resource allocation scheme must be devised which will permit decision making to be more definitive in terms of cost benefit and effectiveness. Essentially, this must involve trade-off analyses among the range of security upgrade options available at each site and their respective ability to more effectively counter the identified threats.

A simplistic view of the dilemma confronting the security system designer is depicted in Figure 1. The application of the classical system engineering process will result in further subsystem definition as system design evolves.

SECURITY SYSTEM DESIGN

Application of the classical system engineering process to the definition and implementation of an integrated security system can be relatively straightforward. However, because of the diversity of threat, vulnerability, configuration and operations of each site, emphasis must be placed upon the input side of the process. For simplicity of discussion in this paper as well as to provide logical decision points in the process, the methodology employed is divided into four successive phases. It should be noted that each site presents unique problems, needs and requirements which can significantly expand one or more of these phases in terms of tasking.

- Phase I—Determine Requirements.
- Phase II—Develop Preliminary System Design.
- Phase III—Develop Final System Design.
- Phase IV—System Implementation.

Each of these phases with their respective steps will be briefly discussed in the remainder of this paper.

Phase I—Determine Requirements

The primary objective of this critical first phase of the process is the determination of vulnerability. It is axiomatic that a security system is only effective insofar as it contributes directly to the reduction and control of vulnerability. Therefore, the process commences with a thorough analysis and definition of threat and the physical and operational environment in which the system must operate. This phase is graphically displayed in Figure 2.

Perform On-site Requirements Analysis. The principal inputs to these first steps are threat and resource documentation.

It is difficult to believe how many sites fail to inventory, assess, update, and thoroughly analyze the threats confronting their operations. As an example, one can find numerous system configurations focused upon an external adversary. The postulated threat is that of an outsider attempting to covertly gain access to the protected area or asset. Little or no consideration has been given to the insider . . . the individual who has authorized access to the asset and/or systems arrayed to protect it. Threat analysis is a key element in all system design considerations and the full range of adversary categories must be clearly postulated. Two sources of data may be utilized for this purpose: generic and empirical based upon actual events. The former will focus upon the nature of the asset or the consequences of penetration or loss and proceed to inventory the adversary characteristics (organization, operations, behavior, resources) which may be presented against the site(s). Actual event data provides more specific information regarding both attempts and successful penetrations of the system in real events. Where no experience exists for the site under review, analogues may be used from similar locations elsewhere. The initial exercise needs to be relatively unconstrained to ensure inclusion of the full range of threats and then progresses to the rejection of lower probability adversary groups and the eventual definition of design basis threat categories.

Resource analysis provides a baseline of in-place or projected subsystem elements which may be considered as contributing to the eventual design. The physical plant,

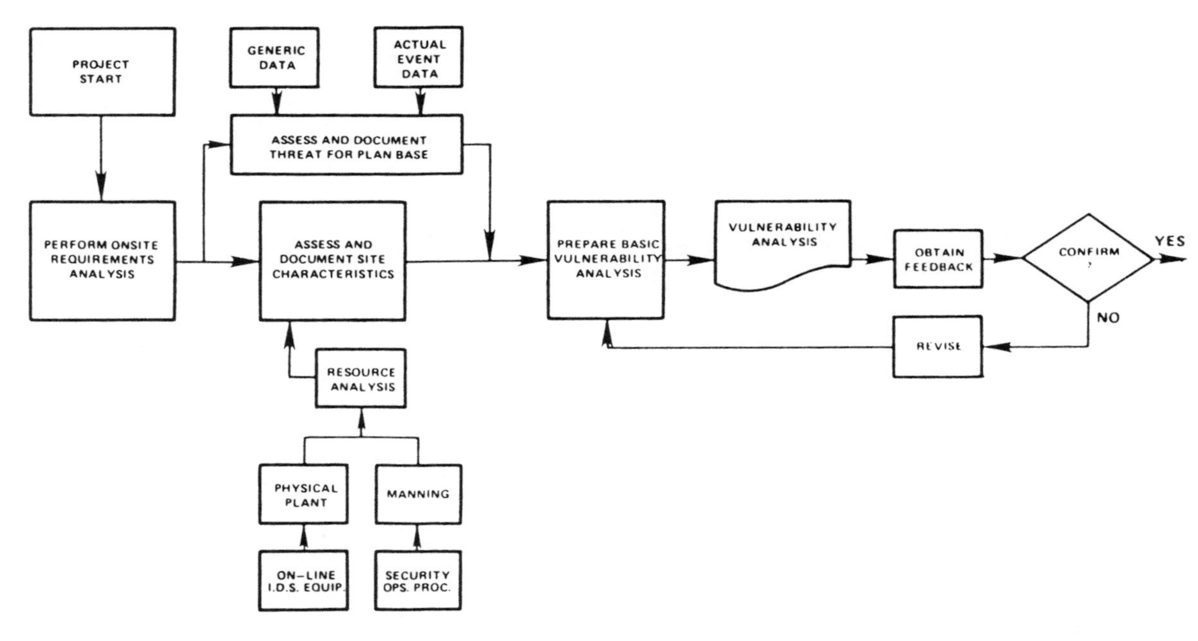

Figure 2. Phase I. Determine Requirements.

for example, has walls, barriers, terrain, etc., with inherent delay capabilities. Various operational procedures are in place which provide for management controls over identified assets. Each of these and other elements must be identified in order to determine their relative capability in terms of protection. This analysis is particularly important in retrofit applications. Cost/benefit/effectiveness alternatives are of critical importance to management and more costly construction and installation options can often be rejected in favor of enhanced procedural controls or staffing. Alternatively, the continuing cost of manpower can be offset through correctly designed electronic subsystems.

In the performance of the resource analysis, an in-depth assessment of the physical environment also needs to be completed. The presence of environmental attributes such as seismic activity (both man-made or natural), radio frequency and electromagnetic interference, weather conditions (rain, snow, fog, etc.), physical conditions of barriers, lighting, terrain features (standing water, hills, ravines, soil, composition, etc.); all will contribute later to potential consideration of electronic sensor siting. Failure to fully evaluate all of these and related factors may completely negate the validity of successive subsystem recommendations.

Preparation of Vulnerability Analysis. Initial steps in the process have generated information on risk (exposure to hazard or loss) and threat (the source of the risk). This information must now be utilized to develop the focal point of the security system design: the vulnerability analysis. Vulnerability may be defined as the relative accessibility of the area or item to be protected to specific risks or threats. As such, successive system design tasks will first determine the range of potential countermeasures which may be employed to remove or control these vulnerabilities and, eventually, lead to a site-specific set of solutions.

The vulnerability analysis breaks out each asset and, given the adversary characteristics established in the design basis threat documentation, proceeds to a determination of the site capabilities to deter, delay, detect and respond to attempts to carry out the postulated adversary sequence. This may be graphically portrayed as seen in Figure 3.

The vulnerability analysis asks the question "what physical and procedural countermeasures must be defeated by the inside/outside adversary to successfully penetrate the protected area, carry out his mission and effectuate an escape." As may be seen, several potential resources may exist to deter, detect or delay his entry at successive points. The outsider, working alone or with other outsiders, is confronted with the full range of subsystems incorporated in the security system. The insider may possess the ability to bypass one or more of the subsystem elements. Insider/outsider collusion threats require the full consideration of redundancy, diversity and the resulting defense-in-depth essential to security system effectiveness.

Consequences analysis is an effective tool in deciding how detailed the subsets of vulnerability assessment need to be for each asset identified. A prioritized matrix of consequences will focus both the user and the system designer on specific vulnerabilities and provide a foundation for later cost-benefit considerations. The resulting analysis should be a focal point for discussion with the user, both to sensitize him to

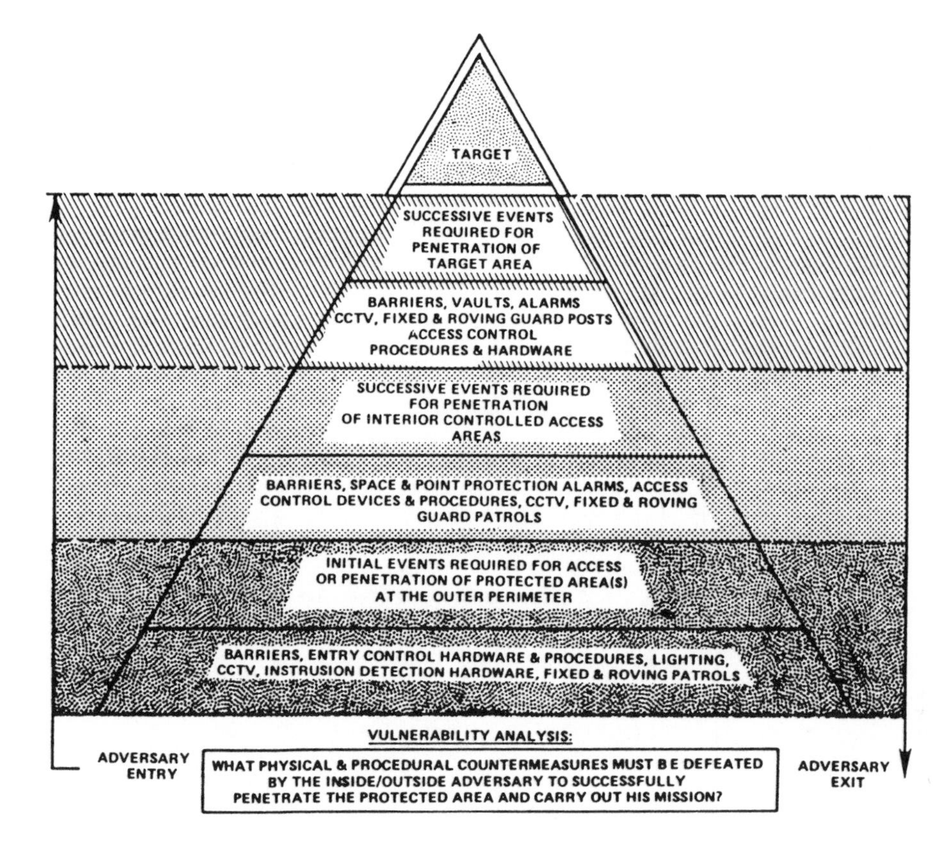

Figure 3. Vulnerability analysis. What physical and procedural countermeasures must be defeated by the inside/outside adversary to successfully penetrate the protected area and carry out his mission?

issues which will require management support later on as well as to obtain consensus for direction in these early phases.

Phase II—Development of Preliminary System Design

Phase II builds upon the essential input data generated in Phase I. The design process is now capable of developing candidate solutions for location and asset-specific vulnerabilities. The steps involved in this phase are shown in Figure 4.

Development of Security Upgrade Plan. The overall system concept is set forth in the upgrade plan. The plan must accommodate the basic functions (mission) of the site while still meeting the objectives of enhanced security. Adjustments and compromises may have to be invoked in considerations regarding work procedures and proc-

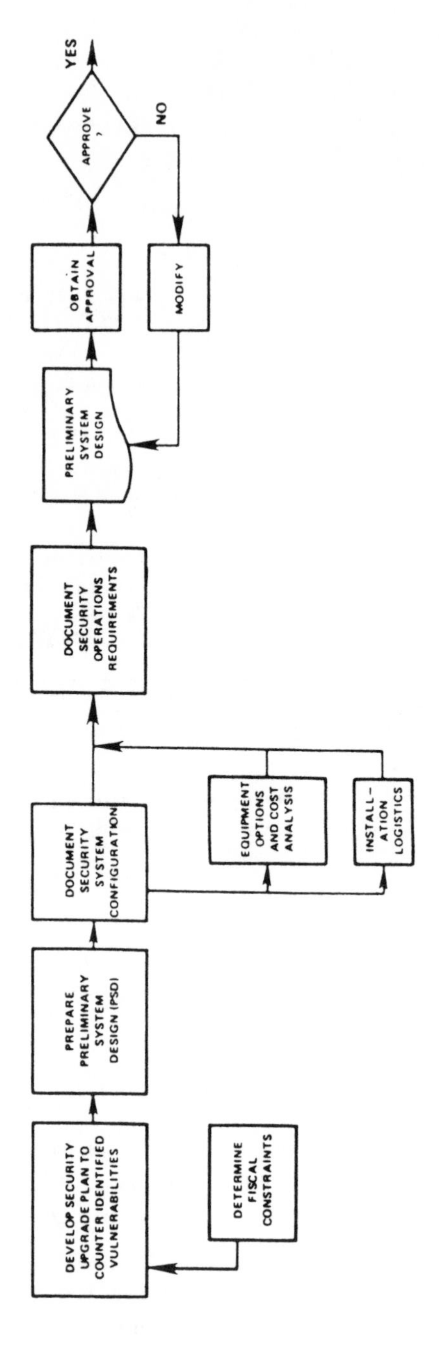

Figure 4. Phase II. Development of preliminary system design.

esses, safety, union agreements, legal/insurance regulations, and other factors and constraints. The design process now commences with the consideration of alternative countermeasures to address each vulnerability. At this point, the principal functional features of terrain, facilities, barriers and sensor subsystems and their performance characteristics are considered. Specific combinations are considered in terms of complementary functions and supplementation of performance capabilities. For example, intruder detection may be provided by above and below ground sensor elements or, physical barriers may be used in conjunction with detection sensors and area surveillance devices. The combinations provide a synergistic effect in burdening the adversary by requiring special equipment and by adding to penetration delay times.

Fiscal constraints are factored into the equation early-on to provide pragmatic limits on design alternatives. Security practitioners must devise a security upgrade plan in consonance with these many constraints and, at the same time, reduce the probability of successful adversary penetrations. The results of the consequences analysis and consensus steps in Phase I can directly serve the interests of good judgment at this juncture.

Preliminary System Design (PSD). The conceptual design proceeds to the preparation of a more definitive security system design. A system configuration complete with identifiable subsystem and subsystem elements can now be set forth. A variety of alternatives may still be available. For example, if the design basis threat is an insider adversary in collusion with outsiders, the mix of procedural, access control and subsystem redundancy features will still be under active debate. Hardware versus personnel intensive solutions may look equally attractive during early stages of PSD development depending upon threat and vulnerability requirements.

A principal set of considerations during developing intrusion detection subsystem discussions will invariably surround the issues of nuisance alarms and probabilities of detection of various electronic sensor configurations. Site-unique environmental data generated during Phase I assessments will provide specific input affecting the candidacy of various phenomenologies and the site preparation required for potential applications. Design basis threat data will directly assist in the determinations of operational requirements for probability of detection success. In many cases, the use of multiple and diverse sensor arrays, logically combined, in conjunction with redundant power and communications capabilities, will provide the necessary defense-in-depth at various zones throughout the protected area.

Cost analysis logistics, both initial and life cycle, involved in system options and the operational enhancement features of alternatives, all assist in the development of a "best" solution to satisfy the unique requirements of the site. The final preparation of the PSD provides the user with a comprehensive overview of:

- how the proposed system will counter the design basis threats identified,
- how the proposed system will mesh with required user operations,
- what policy implications are involved in system implementation,
- what the system (or system options) is likely to cost now and in the future, and
- what consequences he may be able to avoid.

The user can also be confident after preparation of the PSD, that he may proceed with system implementation with assurance that the design represents an optimal solution given the resource commitments approved by decision-makers.

A feedback loop at the end of this phase provides a consensual link for succeeding design steps. With receipt of user approval, the basic system design can be frozen for preparation of the final system design.

Phase III—Preparation of Final System Design

The final system design (FSD) phase sets the security system design process into a pre-implementation cycle and takes the preliminary system design to the levels of specificity required for actual procurement and installation. This phase is seen graphically in Figure 5. It is during this phase that all of the many steps required for subsystem and subsystem element implementation are identified, individual task/event and cost/schedules are prepared, performance specifications to the element level are drawn and required facility renovation/construction plans are developed. A critical element in this phase is the synthesis step of preparing a system integration plan which takes each individual component and specifies its placement, inter-relationship and contribution to overall, integrated system performance. The sum of the parts are specified at this point and the totality can be viewed as an integrated whole.

If this synthesis step could be viewed as a matrix, the horizontal axis would be seen as columns of subsystems. These subsystems are:

- Barrier/Delay subsystem
- Detection subsystem
- Communication subsystem
- Assessment subsystem
- Personnel subsystem
- Equipment subsystem
- Procedures subsystem

The vertical axis would break down each location (or sector) of the site (e.g., outer perimeter zones, clear zone, building perimeter, specific interior locations, etc.). Each column would indicate all subsystem elements required for deterrence, delay, detection and response/operation given the design basis threat(s) postulated in Phase I. An example of the range of options of subsystem elements for the Barrier/Delay subsystem is graphically represented in Figure 6. With the identification of each subsystem element required for the FSD, a system implementation plan can be prepared with supporting documentation. Again, a feedback process is incorporated for end user review and approval.

Phase IV—System Implementation

The final phase is essentially the culmination of plans and specifications emplaced in earlier phases and is displayed in Figure 7. It is most often a real-world test of how

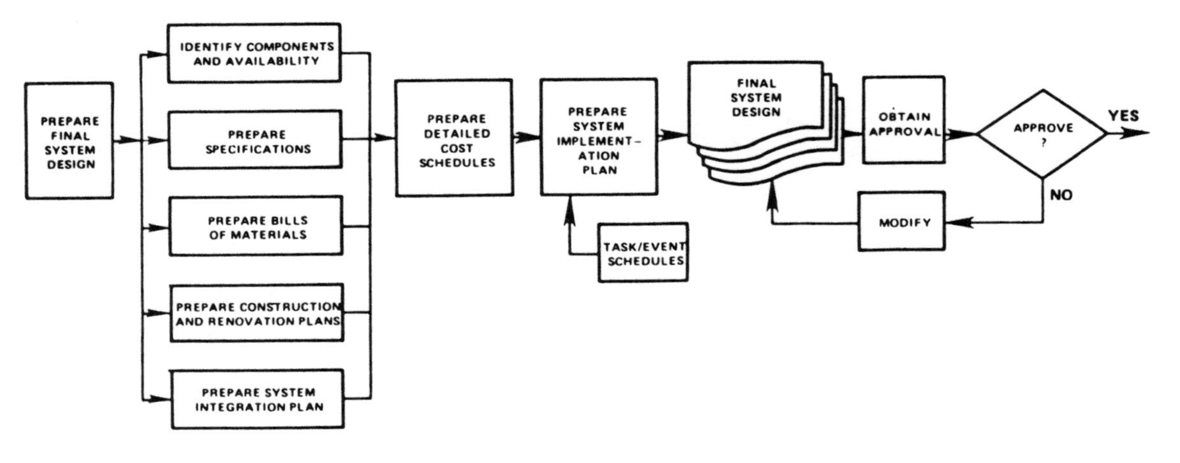

Figure 5. Phase III. Preparation of final system design.

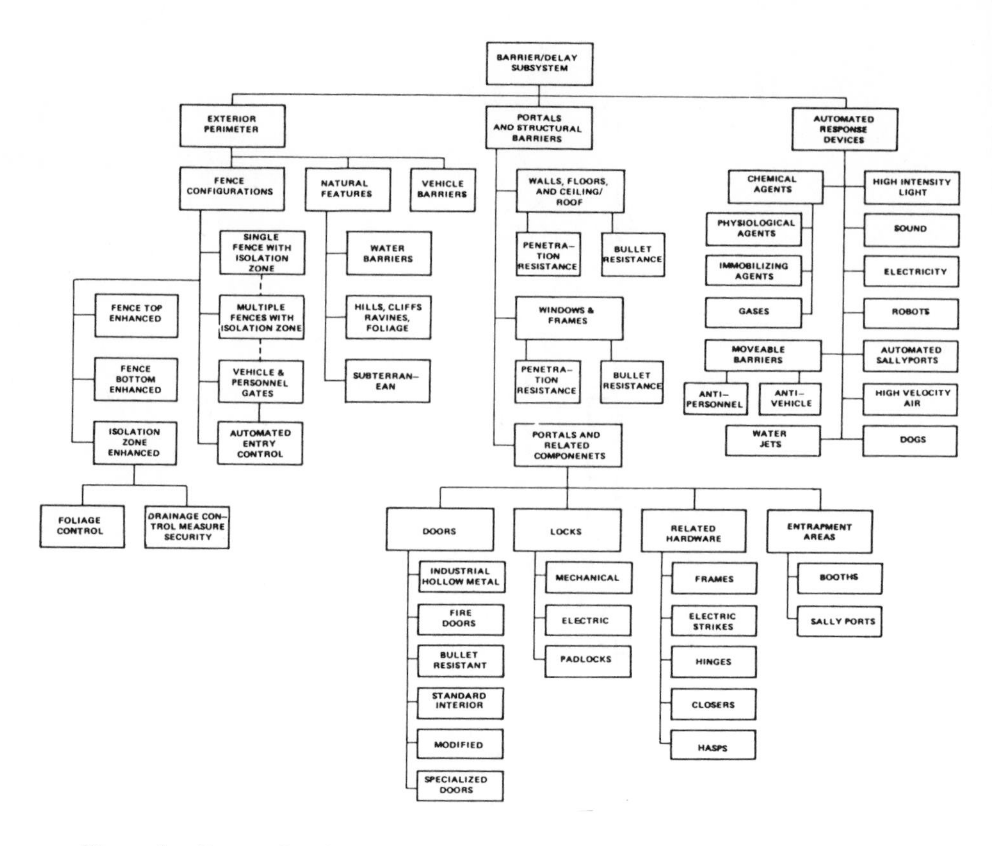

Figure 6. Range of options of subsystem elements for the barrier delay subsystem.

well these earlier phases have been researched, analyzed and documented. The execution of the system implementation plan fully commits monetary resources to the acquisition of hardware and construction/renovation of facilities. Many simultaneous activities are undertaken to procure subsystem components and to prepare essential manuals and procedures for training, operations and maintenance. Receipt, assembly and initial checkout of all capital equipment and miscellaneous hardware is the next essential step. This phase must emphasize the operational requirements of day-to-day system performance by placing concentration upon preparation of all required operations, training and maintenance procedures essential to a fully usable security program. It is absolutely critical that quality control processes be applied throughout this phase. The best system designs and the most effective subsystem elements can fail where installation quality assurance procedures are not followed from the beginning through system checkout and acceptance. Operation and feedback (actual performance in the real world of user operations) will probably be the best test of the system design and implementation phases. Adjustments to procedural routines should be expected throughout the checkout stage. Extensive interviews with operational personnel at all levels should be conducted to define, refine and finalize total system operations requirements.

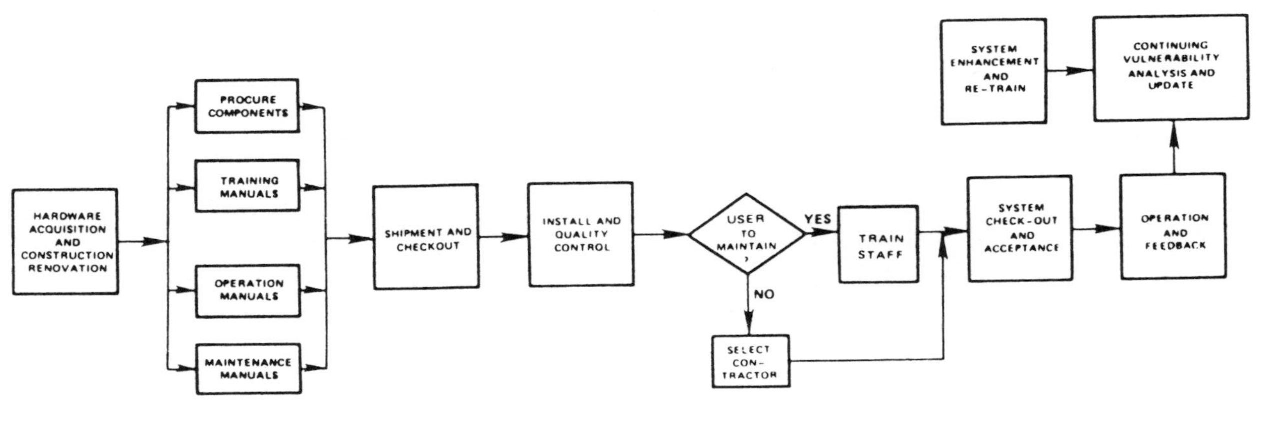

Figure 7. Phase IV. System implementation.

The real and continuing test of the system is capability to meet the design basis threat through periodic vulnerability analysis and update procedures. This involves both overt and covert attempts to penetrate the total system capabilities and reach the protected asset(s). While there are obvious limits on such tests in higher security applications, there is no substitute for an orchestrated test and periodic staff inspection of security capabilities. System enhancement will be the inevitable result of such efforts; they ensure continuing responsiveness to threat and vulnerability dynamics.

CONCLUSIONS

This paper has utilized a classical system engineering process and applied it to the real world of security system design. The ultimate objective of this process is to ensure a totally integrated system design which directly contributes to the reduction of vulnerability with a high degree of reliability and assurance. The array of security subsystems and subsystem elements which comprise an integrated system are depicted in Figure 8. As indicated earlier, it is our empirical observation that, too often, security

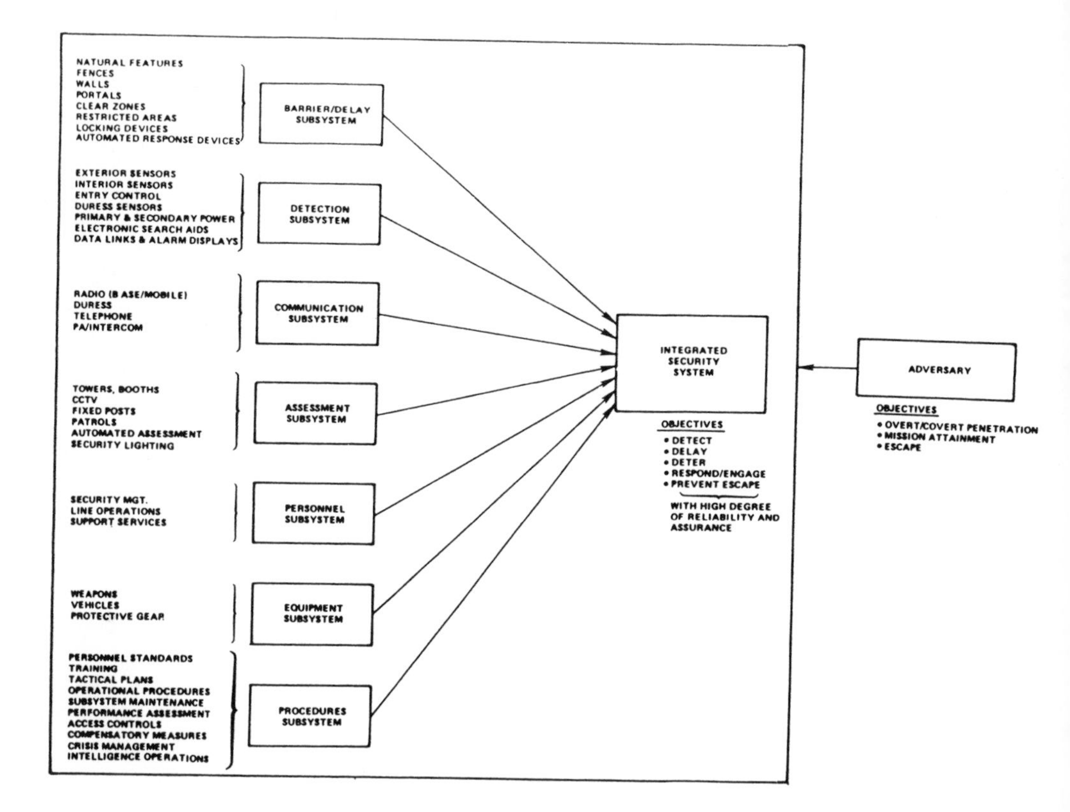

Figure 8. Array of security subsystems and subsystem elements which comprise an integrated system.

system implementation efforts are driven by the pressure of events and/or inadequate attention to vulnerability reduction.

Practitioners in the field often argue that security involves the application of pragmatic lessons gained through experience. Unfortunately, most high security sites cannot afford the lessons of experience. Therefore, in a very real sense, the security system design and implementation process must be approached in a highly organized and thorough manner.

The consequences deserve nothing less.

PART II

Security Barrier Technology

Typically, barriers are viewed by security administrators as necessary but unexciting, stationary structures. Present advances in barrier technology, however, will change this view. The U.S. Department of Defense and the State Department are encouraging and supporting research in barrier design. Much of this interest is generated by the 1983 attack on the Marine Headquarters in Beirut in which a truck heavily laden with explosives penetrated the perimeter to kill 241 people. The forthcoming reports will be valuable additions to subsequent books.

It is doubtful, however, that any barrier could be more exotic than the barriers discussed in this section. Active barriers, the subject of this section, certainly do not fit the stereotype of the static structures of a fixed barrier. They are highly dynamic, complex and sophisticated.

Fixed barriers, especially perimeter barriers, only give an illusion of security. The classic 8-foot chain link fence topped by barbed wire can be scaled in a matter of seconds. Pairs of perimeter barriers and layered intrusion detection devices may increase the delay. Yet, if the target is extremely critical and the presumed threat is sufficiently equipped, traditional security technology may be insufficient. Active barriers as presented here can increase the delay time to 30 minutes, providing valuable time to apprehend the adversary.

The active barrier system discussed here was designed and installed in facilities involved in national welfare. The two articles are developmental in nature. Yet the system is now readily available for commercial uses such as the protection of safety deposit departments of banks or critical areas of museums. The system is beneficial anywhere there are highly valuable or critical items contained in a confined space.

The first article addresses the nature of the activated barrier system and the second article progresses from the first and deals with procedures for deploying the active barrier. The authors stress the desirability of employing human security operators in the deployment function, but stress the benefits of logic and computer assistance in the decision process. The system could be totally computer operated in lieu of on duty security personnel.

The two articles combined show the synergistic value of a system composed of fixed barriers, intrusion detection devices, access control devices, closed-circuit television, computers, policy, procedures and personnel in deploying the active barriers to effectively thwart even the most sophisticated and determined adversary.

Physical Protection System Using Activated Barriers

Ronald E. Timm
Thomas E. Zinneman
Joseph R. Haumann
Hayden A. Flaugher
Donald L. Reigle

Abstract.The Argonne National Laboratory has recently installed an activated barrier, the Access Denial System, to upgrade its security. The technology of this system was developed in the late 70s by Sandia National Laboratory-Albuquerque. The Argonne National Laboratory is the first Department of Energy facility to use this device. Recent advancements in electronic components provide the total system support that makes the use of an activated barrier viable and desirable.

Typically, well-designed fixed barriers provide delays on the order of minutes to multiple, sophisticated adversaries. An equally well-designed activated barrier will improve the delay by an order of magnitude for the same threat. Further, it is desirable that the effects of an activated barrier be benign for equipment and personnel in the vital area.

The premise of an activated barrier is that it is deployed after a positive detection of an adversary is made and before the adversary can penetrate a vital area. To accomplish this detection, sophisticated alarms, assessment, and communications must be integrated into a system that permits a security inspector to make a positive evaluation and to activate the barrier. The alarm sensor locations are selected to provide protection in depth. Closed circuit television is used with components that permit multiple video frames to be stored for automated, priority-based playback to the security inspector. Further, algorithms permit "look-ahead" surveillance of vital areas so that the security inspector can activate the Access Denial System in a timely manner and not be restricted to following the adversaries' penetration path(s).

1984 Carnahan Conference on Security Technology, University of Kentucky, Lexington, Kentucky, May 16–18, 1984.

A physical protection system utilizing activated barriers can provide the delay necessary so that initial response can be accomplished by a smaller force. Hardening of vital areas can also be accomplished in a more cost-effective manner through the complementary use of an access denial system.

INTRODUCTION

A basic physical protection system has three functional subsystems: Detection, Delay, and Response. The threat spectrum challenging these subsystems has sophisticated weaponry and techniques with personnel dedicated to the successful completion of their mission. The defeat of any one of the subsystems in a basic system is sufficient for the adversaries to successfully meet their objectives.

Advancements in detection technology have improved the efficiency of sensor components so that the probability of detection is nearly unity; layering of the sensors virtually assures detection. The improved training and equipping of the response force, particularly the addition of SWAT teams, ensures successful confrontation of the adversaries.

Delay is the security feature that ties together detection and response to defeat the adversary. In the past, barriers consisted principally of fixed delays, e.g., doors, walls, and fences; improvements consisted of bigger, thicker, heavier, and more. Since the mid-70s, research has been under way to provide a new type of barrier—an active barrier.[1] An active barrier is a component that is installed in a passive mode and is activated by a security inspector after the positive assessment of an adversary. When the activated barrier is deployed, a delay of 30 minutes or more is provided.

PHYSICAL PROTECTION

Physical protection systems are often designed using a layered technique. Figure 1 shows a typical layered security system that provides protection in depth for a vital area. The ''yellow'' zone is typically the perimeter of a Protected Area (PA); the orange zone is an intermediate zone, usually within a building; the red zone is directly outside a vital area and may include the Materials Access Area (MAA). Each zone has layered sensors of perimeter and volumetric alarms, with closed circuit television (CCTV) for assessment. As can be seen in Figure 1, the yellow zone provides the initial detection for ground penetration, and the orange zone provides the first level of detection for airborne penetration. Further, the yellow zone, because of adverse environmental conditions, has a probability of detection considerably less than unity, and because it is the farthest out, with the longest perimeter, it is the most costly to construct. Therefore, the outside PA is the weakest detection point and the most costly to deploy. As the detection zone moves nearer to the vital area, it usually moves indoors, where it is environmentally less hostile, the perimeter becomes shorter, and detection is less costly. The inner defense zones are more easily controlled.

In the recent past, the layering technique, with emphasis on the Protected Area, was necessary because, after detection of the adversary, delay was provided by distance

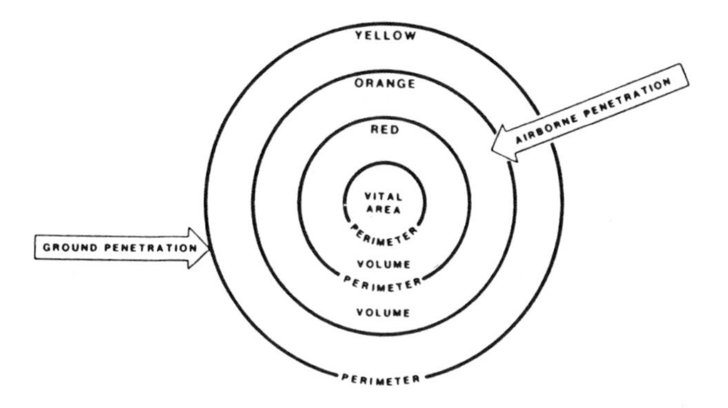

Figure 1. Protection in depth.

to travel plus a series of fixed barriers. Figure 2 shows a typical penetration diagram through the defense zones to a vital area for either an act of theft or sabotage. The fixed barrier system is only as good as it is deep. Improvement in distance was often not possible, and improvement in hardening barriers is expensive to implement. Further, the addition of fixed barriers often adversely impacted operations. With the advent of active barriers, a new option is opened. An active barrier can be located in a vital area, and, once it is deployed, can provide a delay of 30 minutes or more. Thus, the detection point shifts dramatically to the orange or red zones, where probability of alarm and assessment improves virtually to unity in all weather.

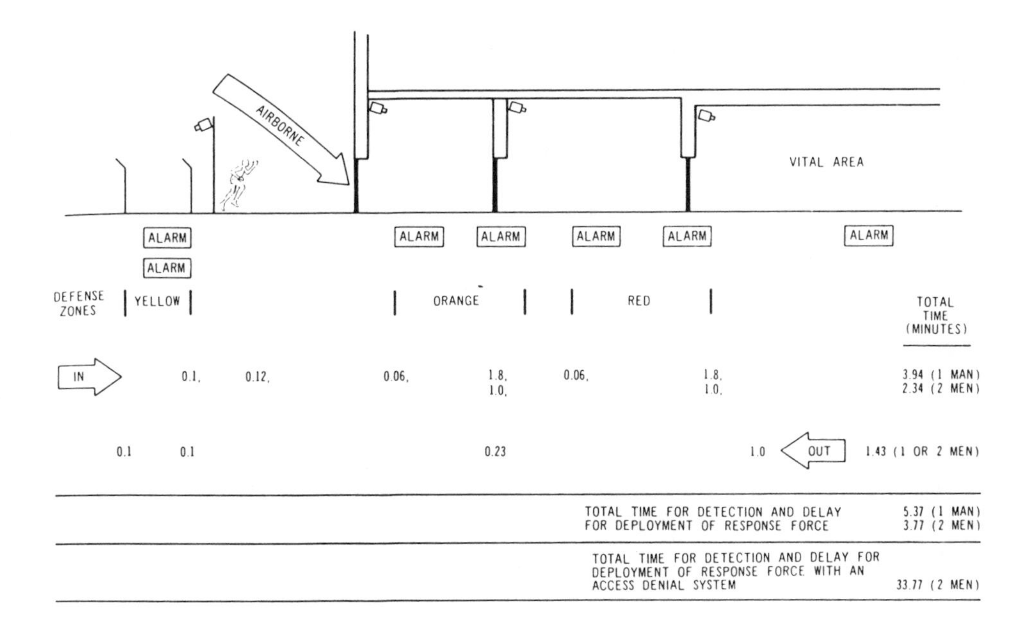

Figure 2. Penetration diagram.

BARRIER TECHNOLOGY

Barrier technology is divided into two classes—fixed or activated. The fixed barriers of doors, walls, etc., were briefly discussed above. The active barriers fall into four generic classes; cold smoke, aqueous foam, rigid foam, and sticky foam. Table 1 shows a matrix of characteristics. One of the most interesting aspects of the activated barriers is its effect on personnel and equipment in the vital area, if the barrier is deployed. Two of the activated barriers, rigid and sticky foam, are considered bad for personnel and equipment. The remaining two, aqueous foam and chemical smoke, are benign; in fact, neither are harmful to personnel, even if they are exposed throughout the persistancy of the deployment, and both are also not harmful to mechanical or electronics components. Total clean-up time after deployment is on the order of a week.

Activated barriers are typically used with physical restraints. The activated barriers, together with the physical restraints, comprise an Access Denial System. The physical restraints are placed over and around the sensitive items in the vital area. Their purpose is to stop an adversary from grabbing and snatching a sensitive item even though an active barrier is, or has been, deployed. The synergistic effects on adversary movement in an active barrier environment, plus adversary action with a fixed restraint, means that a delay of 30 minutes can be achieved for an Access Denial System. Table 2 shows the relative effectiveness of fixed barriers and an access denial system.

The deployment of an activated barrier requires: (1) the security inspector ARM and FIRE operation sequence, (2) barrier deployment time, and (3) barrier sustaining time. Obviously, the key to success for an activated barrier is the security inspector operation sequence. Additionally, the activated barrier equipment must be hardened to ensure its successful deployment. Since the equipment is located in the vital area, its hardening must be equivalent to that of the vital area perimeter, or approximately one minute for multiple adversaries in the red zone.

The cost of an Access Denial System, activated barrier and restraints, is on the order of $200,000. By almost any component cost comparison, this is cost effective. A simple cost comparison of fixed barriers vs an ADS for the same delay yields cost savings of greater than five to one. Further savings are effected by reduced demands

Table 1 Active Barrier Selection Matrix

	Cold Smoke	*Aqueous Foam*	*Rigid Foam*	*Sticky Foam*
Persistency	Good	Marginal	Good	Good
Premature consequences	Good	Good	Bad	Bad
Safety	Good	Good	Marginal	Bad
Volume	Large	Large	Small	Small
Cost	Good	Good	Marginal	Marginal
Availability	Commercial	R&D	R&D	R&D

Table 2 Typical Barrier Delays (Minutes)[2]

	Adversary(s)	
	1 man	*2 men*
Fixed Barriers		
Fences	0.12	—
Walls—8″		
Block	0.8	—
Reinforced	—	1.9
Doors—Industrial		
Standard	0.2	—
Hardened	1.8	1.0
Distance (100 ft)	0.12	0.12
Activated Barrier (Smoke)		
ARM/FIRE	0.2	0.2
Deploy	1.1/1.6	1.1/1.6
Sustain	28.1/34.0	28.0
Restraint/Defeat	1.5	0.5

on the response force; personnel can be assigned normal duties and emergency duties, rather than to one, exclusively.

BARRIER OPERATION

Barrier operation consists of the successful detection of an adversary and successful security inspector ARM and FIRE operation sequence.

Successful detection of an adversary is achieved by the combination of two separate operation scenarios using the same detection equipment. The first operation sequence requires acknowledgment of alarms of ascending priorities (yellow, orange, and red), and their successful assessment using state-of-the-art CCTV technology. The modern CCTV technology consists of cameras and video switchers that are interfaced directly to the alarm processors. The output of the switchers provides real-time video to the security inspector; the TV monitors and stores discrete frames of video in memory devices. The stored images are played back for assessment simultaneously with the real-time video.

The tracking of adversary action is one operation sequence; the second operation sequence involves using the same CCTV system to survey the vital area for the coming of the adversary(s). This allows the security inspector to "look ahead" and to activate the barrier before the adversary(s) begins the attempt to enter the vital area. A vital area can have up to six exposed faces that an adversary could choose to penetrate. The most vulnerable face(s) would be the entrance(s). To scroll through the video of the six faces is not operationally effective, particularly in an installation having multiple activated barriers. Therefore, the "look ahead" video is displayed only for the entrance of a vital area until such time that an alarm is detected on any of the remaining five faces. With an alarm at the vital area, alternate cameras covering the entrance and the alarmed zone will be presented to the security inspector for assessment.

Assessment and surveillance require the action of two security inspectors to maximize the effectiveness of the active barrier. In most instances, high security installations have a central alarm station (CAS) and a secondary alarm station (SAS), each with one security inspector. Procedures have to be developed whereby the CAS security inspector has the responsibility of assessing the adversary(s), and the SAS security inspector has the responsibility for the "look ahead" surveillance of the adversary. In an emergency, either security inspector is capable of performing both the assessment and surveillance functions.

The joint operations action also lends itself to additional hardening of the activated barrier; that is, the barrier can be activated from either of the two locations, the CAS *or* the SAS. Further hardening of the activated barrier can be achieved by deployment of an activated barrier that will "fire or fail" when all control lines are deactivated.

CONCLUSIONS

Activated barriers have been under development since the mid-70s, but it is only with the recent advances in detection technology that they have become a viable component in a well-designed physical protection system. With the use of layered alarm sensors monitored by high-speed computer processors and integrated with modern closed circuit television, security inspectors are able to acknowledge and assess alarms accurately in tens of seconds. With timely information, a hardened activated barrier improves the delay time from minutes to tens of minutes. Another security inspector in a diverse location, using surveillance monitors in a complementary mode, enhances the overall system effectiveness.

The successful delay of the adversary with the use of an Access Denial System means that the response force can provide an initial response with fewer personnel and have more time to deploy a final force to meet multiple, sophisticated adversaries. The use of activated barriers and restraints increases the response force deployment time from minutes to tens of minutes, particularly for theft scenarios. Even for sabotage, the time for a response force to effect action has the same order of magnitude improvement.

An Access Denial System (activated barrier and restraints), compared to fixed barriers for the same delay, yields cost savings of greater than five to one, with additional savings effected by reduced demands on the response force.

REFERENCES

1. John W. Kane and Martin R. Kodlick, "Access Denial Systems: Interaction of Delay Elements," *Proceedings, 24th Annual Meeting on Nuclear Materials Man-*

agement, vol. 12 (Vail, Colorado: Institute of Nuclear Materials Management, July 1983), pp. 301–6.

2. *Barrier Technology Handbook*, Sandia National Laboratories-Albuquerque, SAND77-0777, rev. ed., 1981.

Developing Deployment Procedures for Activated Barriers

Ronald R. Rudolph
Ronald E. Timm

Abstract. As physical protection systems are upgraded to use state-of-the-art components, decision-making processes facing system operators become more complex. Traditionally, protection systems consist of alarm sensors and processors for detection, and closed-circuit television systems for assessment and surveillance. Recently, activated barriers have been introduced in high-security applications. Deployment of an operator-activated barrier can delay an adversary 30 or more minutes. In order to be effective, these systems require timely and accurate assessments. The time available for the operator to complete the assessment process and to decide whether to deploy the system may be quite short, less than 3 minutes. Because an activated barrier can be deployed only once, the assessment must be accurate. If the activated barrier is deployed when no threat exists or after a successful adversary challenge, it is ineffective and costly. We have studied the various options facing a security-system operator in the course of normal operations (when a security area is in the access or secure mode), and during abnormal conditions (e.g., emergencies). The results provide a methodology for developing procedures that the operator can use to ensure successful and timely barrier deployment.

INTRODUCTION

Effective use of operator-activated barriers (cold smoke, aqueous foam) as part of a high-security physical protection system (PPS) requires explicit and readily implemented deployment procedures. Activated barriers may be installed along access paths

1985 Carnahan Conference on Security Technology, University of Kentucky, Lexington, Kentucky, May 15–17, 1985.

to or inside of the target area.[1] The example given here is for the latter type of installation in a generic PPS, where this system is the last barrier to delay unauthorized access to the target. Similar procedures can be developed for choke-point applications along access paths.

The decision to deploy the activated barrier is usually based on an accurate assessment of alarms. Having completed an assessment, the operator follows established barrier-deployment procedures that maximize the protection provided the target, while minimizing the possibility of an incorrect deployment. In some cases, the decision may be made with incomplete or no assessment. For most facilities, the complexity of the PPS and the number of possible inputs to the deployment decision make it necessary to carry out a systematic analysis of all possible input combinations, before a set of deployment procedures can be established. The deployment procedures should use simple logic decisions. The alarm system process computer could be programmed to make decisions that are based solely on alarm states. However, most of the decisions use assessments as inputs. Therefore, operator control of the activated barrier is always recommended. The development of deployment procedures identified a need for the assessment systems near and in the target area to have a ''look-ahead'' capability to track and to anticipate adversary movement.

ASSUMPTIONS

In a high-security system, the target area (e.g., Vital Area) is surrounded by concentric zones of three-dimensional protection (detection and delay), which we will call the Red Zone (e.g., the Materials Access Area), the Orange Zone (e.g., the building boundary), and the Yellow Zone (e.g., the Protected Area) (see Figure 1 for a two-dimensional representation of protection zones). If the level of protection provided within each zone is balanced, an adversary must pass through and be detected in each, before reaching the target area. Note that this does not hold for the Yellow Zone in Figure 1, which represents an exterior zone that can be circumvented by an airborne penetration, by sophisticated bridging techniques, or by waiting for inclement weather.

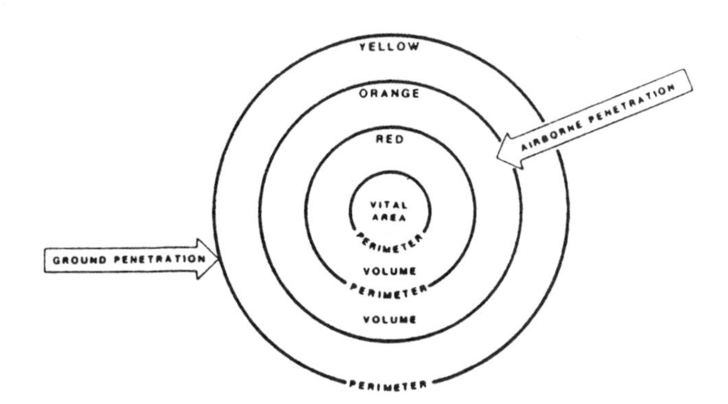

Figure 1. Physical protection zone.

Detection in a zone is defined to comprise alarm sensor and assessment systems, and detection procedures. An alarm from a sensor system alerts the operator to an abnormal situation in a zone. Alarms may come from sensors within the volume of the zone and/or from perimeter sensors outside each barrier. (Sensors precede barriers, to allow for delay during assessment.) Assessment systems and procedures are used to evaluate the alarm. An appraisal of the PPS status, based on these assessments, is then used to make a decision to deploy an activated barrier. The decision may be based solely on sensor alarms; however, a decision can not be based solely on an assessment without any alarm, i.e., no credit will be taken for surveillance procedures to initiate an alarm. Once an alarm has been received, "look-ahead" assessment procedures for the Red Zone and target area can be used to track and to anticipate adversary movement.

Sensor Systems

Each protection zone should contain components to successfully sense a penetration attempt. To maximize alarm reliability, redundancy and diversity of high-quality and complementary components should be used. A zone will be in one of three possible sensor states, at any time:

- Secure—all sensors operative with no alarms;
- Access—sensor(s) disabled for authorized activity or maintenance, or degraded sensor capability in zone; and
- Alarm—real, nuisance, or false alarm(s) generated.

The sensing capability of a protection zone is considered to be degraded at any time that a sensor in the zone is not operating as specified. In addition to component degradation, this might be caused by weather conditions or by external interferences (e.g., EMI). The sensor state for a zone is determined by the lowest state (in the order above) of all sensors in the zone.

Assessment Systems

An assessment of the situation in a zone is based on all information received from detection systems and procedures. An assessment is defined to be positive if an alarm is received from a sensor and there is sufficient information available for the operator to conclude that unauthorized activity has occurred, is in progress, or is imminent. An assessment is defined to be negative if an alarm is received from a sensor, but sufficient information is available for the operator to conclude that unauthorized activity is not imminent or in progress, i.e., all assessment systems are operative and no unauthorized activity can be monitored.

Inaccurate positive assessments, caused by marginal equipment, lower the confidence of security personnel in the PPS and may result in incorrect deployment of the activated barrier. Inaccurate negative assessments, caused by incomplete or erroneous information, weaken or negate the effectiveness of the PPS, endangering the target. Both of these consequences severely reduce the effectiveness of the PPS.

Each protection zone should contain components to accurately assess alarms. A zone will be in one of three possible assessment states, at any time:

- Negative—no unauthorized activity is evident;
- Inoperative—disabled assessment system, or degraded assessment capability; and
- Positive—unauthorized activity has occurred, is in progress, or is imminent.

The assessment capability of a protection zone is considered to be degraded at any time that a definite negative assessment of the situation in the entire zone can not be made. In addition to component degradation, this can be caused by weather conditions or by temporary physical obstructions. The state of the assessment system in a zone is determined by the lowest state (in the order above) of all assessment components in the zone.

Sensor/Assessment System Combinations

Based on all possible states for sensor and assessment systems, there may be up to six possible sensor/assessment system combinations in a protection zone: Secure/negative; Secure/inoperative; access; alarm/negative; alarm/inoperative; and alarm/positive. The combination secure/positive is not viable, because it implies that assessment can precede sensing, which by an earlier assumption is not allowed. Similarly, if the sensor systems in a zone are in access, the assessment system can only be used for surveillance and the assessment state is irrelevant.

The level of protection provided in a zone is equal to the minimum protection provided by barriers and detection systems in the zone. A vulnerability analysis for the facility will provide an estimate of the delay time for each barrier and identify potential penetration paths. The level of protection provided by different zones need not be the same. For our example, each zone has volume and perimeter penetration sensors, and assessment components. Minimum penetration delay times for each barrier and volume in our example facility (Figure 2) are typical.[2] When developing deployment procedures for activated barriers, penetration-delay times must be compared with assessment and barrier-deployment times, to ensure target protection.

Activated Barrier Operation

An activated barrier has three operating modes: standby, arm, fire. The activated barrier is normally in the standby mode. The deployment sequence is to arm, and then to fire. The system can be placed in the armed mode indefinitely, while an alarm is assessed further. If the system has been armed and a negative assessment determines that it is no longer needed, it can be returned to the standby mode. Once the operator has sent the fire order, there is a user-set delay period for area evacuation before the

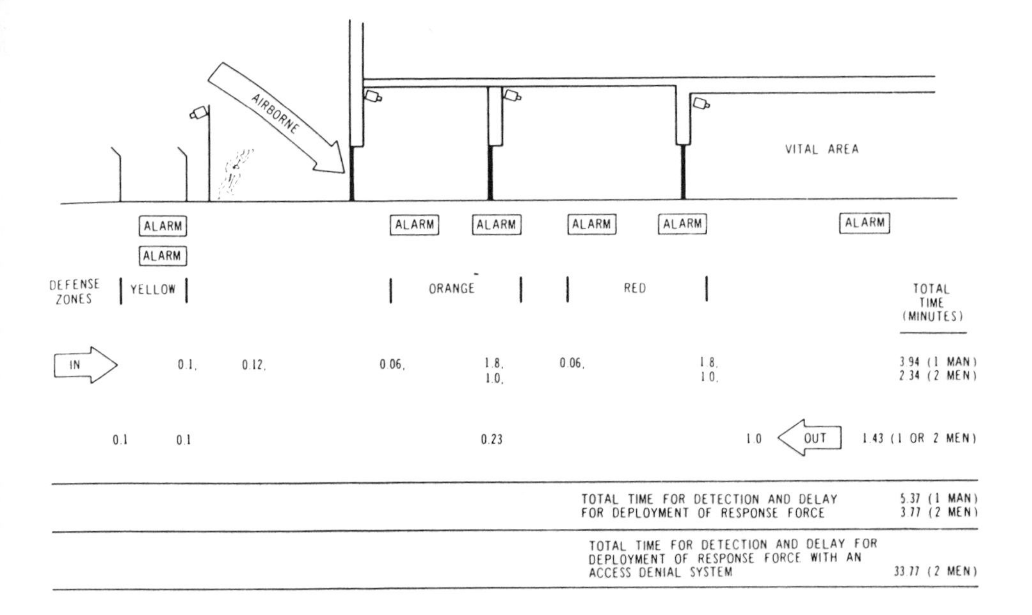

Figure 2. Typical penetration path.

system deploys. For our example, this will be 0.25 minutes. During this time period, the system can be returned to the arm or standby modes safely. There is a time period after the system is fired until the coverage provided by the delay medium is sufficient to protect the target. This will be 1.5 minutes, for our example. Because of this coverage delay, assessment is of little value in the target area, except to monitor the state of the activated barrier system. Once deployed, the activated barrier (when combined with physical restraints) can delay an adversary for 30 or more minutes.

At any time, there are three possible responses for the operator to an alarm:

- None—leave activated barrier system in current mode (standby, arm, fire);
- Arm the system; or
- Fire the system.

While developing deployment procedures for activated barriers, we identified the need for "look-ahead" assessment capabilities in the Red Zone and in the target area. The time available for the barrier operator to complete the assessment process and to decide whether to deploy the system may be quite short, less than 3 minutes. Therefore, the operator should be provided a dedicated assessment monitor (ideally located on the display panel for the activated-barrier system), whose display follows a predetermined algorithm. For our example, the algorithm will be

- No alarm—real-time image of target area entrance from outside;
- One alarm—alternate between real-time image of target area entrance from outside and frame-grabber image from alarm assessment camera at time of alarm; and

- Multiple alarms—alternate between real-time image of target area entrance from outside and real-time image of interior of target area floor, ceiling, and walls (other than entrance).

The dedicated "look-ahead" feature allows the operator to make a better assessment of an alarm and to more effectively deploy the activated barrier, if needed. The use of this feature requires two operators: one to follow the path of the adversary penetration and one to watch the target area. Such procedures will nullify the negative assessment state for alarms in access, inoperative assessment, and diversionary tactics.

METHODOLOGY

Logic Tree

The approach in developing deployment procedures for an activated barrier is to identify all possible sensor/assessment system combinations for the PPS, using a logic-tree type of analysis. Part of the reduced tree for our generic PPS, when the target area sensors are secure, is shown in Table 1. The other part, when the target area sensors are in access, is shown in Table 2. Each branch of the tree represents one possible combination of states of the sensor and assessment systems in a protection zone. All branches for the same protection zone are in the same column. Thus, moving from the left to the right through the tree corresponds to moving out from the target through layers of detection. For our example, the Yellow Zone has been left off the tree.

A leg through the tree is defined as a series of branches from the target through the protection zones. Each branch along the leg provides the status information about another protection zone. The leg is, therefore, a picture of the status of the entire PPS at a single point in time. For every such leg, a decision must be made as to whether the mode of the activated-barrier system should be modified.

With two exceptions, the sensor/assessment system states in Tables 1 and 2 are as defined earlier. First, the "look-ahead" capabilities in the target area and Red Zone are reflected in the assessment state "Look," which is only allowed when there is an alarm in the Red and/or Orange Zones. Second, when the assessment system in a zone does not provide enough information to make a positive assessment, the state of both sensor alarms (volume and perimeter) may become the determining factor in the operator response.

Normal Operations

Activated-barrier deployment procedures should allow for normal access to and egress from the target and surrounding areas. The legs of the tree that are followed for normal access of authorized personnel to the unoccupied target area are identified in Tables 1 and 2 (legs 1, 3, and 12), with leg 1 being the initial state of the system, when it is the most secure. The final state for any access to the target area is leg 40 in Table

Table 1 Activated-barrier Decision Tree (Target Area Secure)

Red Zone	*Orange Zone*	*Leg Number*	*Normal Access*	*Normal Maintenance*	*Operator Response*
Secure/Neg.	Secure/Neg.	1	*		
	Secure/Inop.	2		*	
	Access	3	*	*	
	Alarm/Neg.	4			Arm
	Alarm/Inop.	5			Arm
	Alarm/Pos.	6			Arm
Secure/Inop.	Secure/Neg.	7		*	
	Secure/Inop.	8			Proc.
	Access	9			Proc.
	Alarm/*****	10			Arm
Secure/Look	Alarm/*****	11			Fire
Access	Secure/Neg.	12	*	*	
	Secure/Inop.	13			Proc.
	Access	14			Proc.
	1 Alarm/Neg.	15			Arm
	2 Alarms/Neg.	16			Fire
	1 Alarm/Inop.	17			Arm
	2 Alarms/Inop.	18			Fire
	Alarm/Pos.	19			Fire
Access/Look	Alarm/*****	20			Fire
1 Alarm/Neg.	Secure/Neg.	21			Arm
	Secure/Inop.	22			Arm
	Access	23			Arm
	1 Alarm/Neg.	24			Arm
	2 Alarms/Neg.	25			Fire
	1 Alarm/Inop.	26			Arm
	2 Alarms/Inop.	27			Fire
	Alarm/Pos.	28			Fire
2 Alarms/Neg.	**************	29			Fire
1 Alarm/Inop.	Secure/Neg.	30			Arm
	Secure/Inop.	31			Arm
	Access	32			Arm
	1 Alarm/Neg.	33			Arm
	2 Alarms/Neg.	34			Fire
	1 Alarm/Inop.	35			Arm
	2 Alarms/Inop.	36			Fire
	Alarm/Pos.	37			Fire
2 Alarms/Inop.		38			Fire
Alarm/Pos.		39			Fire

Table 2 Activated-barrier Decision Tree (Target Area in Access)

Red Zone	*Orange Zone*	*Leg Number*	*Access Leg*	*Maintenance Leg*	*Operator Response*
Secure/Neg.	Secure/Neg.	40	*		
	Secure/Inop.	41			Proc.
	Access	42	*	*	
	Alarm/Neg.	43			Arm
	Alarm/Inop.	44			Arm
	Alarm/Pos.	45			Fire
Secure/Inop.	Secure/Neg.	46			Proc.
	Secure/Inop.	47			Proc.
	Access	48			Proc.
	Alarm/Neg.	49			Arm
	Alarm/Inop.	50			Arm
	Alarm/Pos.	51			Fire
Secure/Look	Alarm/*****	52			Fire
Access	Secure/Neg.	53	*	*	
	Secure/Inop.	54			Proc.
	Access	55			Proc.
	1 Alarm/Neg.	56			Arm
	2 Alarms/Neg.	57			Fire
	1 Alarm/Inop.	58			Arm
	2 Alarms/Inop.	59			Fire
	Alarm/Pos.	60			Fire
Access/Look	Alarm/*****	61			Fire
1 Alarm/Neg.	Secure/Neg.	62			Arm
	Secure/Inop.	63			Arm
	Access	64			Arm
	Alarm/*****	65			Fire
2 Alarms/Neg.	**************	66			Fire
1 Alarm/Inop.	Secure/Neg.	67			Arm
	Secure/Inop.	68			Arm
	Access	69			Fire
	Alarm/*****	70			Fire
2 Alarms/Inop.	**************	71			Fire
Alarm/Pos.	**************	72			Fire

2. The legs of the tree that are followed for a normal access of authorized personnel to the target area, when it is already occupied, are identified in Table 2 (legs 40, 42, and 53), with leg 40 being the initial and final states of the system.

Activated-barrier deployment procedures should also allow for normal maintenance of sensor and assessment systems, while sustaining a high level of protection. In the tables, normal maintenance for an assessment system is contained within the inoperative state for the system; maintenance for a sensor system is contained within the access state. The legs of the tree that include normal maintenance are also identified (legs 2, 3, 7, 8, 42, and 53).

Activated Barrier Deployment

Operator responses to the other legs of the logic tree are shown in the last column of each table. In addition to the arm and fire responses, there are situations that should either be prevented by operating procedures (e.g., limit access in the target area to one zone at a time) or corrected immediately (e.g., use temporary guard patrols to cover areas with degraded detection components). These are minimum responses, based on the assumptions above. They can be changed to provide a more conservative system, reducing the complexity of the logic decisions. However, such changes would increase the probability of barrier deployment when not required.

For a particular facility, deployment procedures for each leg of the tree would be based on the level of protection provided by each zone (delay times and detection systems), as well as the response times of the operator and of the activated barrier.

CONCLUSIONS

Using a set of logic trees, deployment procedures can be developed for an activated barrier in a facility that has a high-security PPS. These procedures allow for normal operations, while providing the maximum protection for the target. Key to the implementation of the procedures is the use of a dedicated "look-ahead" assessment system, as part of the display panel for the activated-barrier system.

REFERENCES

1. R. E. Timm et al., "Physical Protection System Using Activated Barriers," *Proceedings, 1984 Carnahan Conference on Security Technology.*
2. *Barrier Technology Handbook*, Sandia National Laboratories—Albuquerque, SAND77-0777, rev. ed., 1981.

PART III

Intrusion Detection Systems

A facility is defended by means of perimeter and internal intrusion detection systems. These systems provide lines of defense against possible intrusions. Perimeter barriers constitute the first line of defense. Because violation of perimeter barriers can occur in a matter of seconds, devices are needed to reinforce resistance to intruders. Perimeter intrusion detection devices are the means to shore up the defenses of facilities. Intrusion detection systems are comprised of four components: sensors, a communication link, controls and annouciators (or alarms).

The first two articles in this section discuss qualities of one currently popular perimeter intrusion detection sensor: ported coaxial cable. The articles discuss the qualities of the sensors and the burial medium effects of ported or leaky coaxial cable sensors, as well as compare long-line and block-sensor designs.

The concept of the ported coaxial cable as a perimeter intrusion sensor was first introduced by Dr. Keith Harman, the author of the first article, and a co-author in 1976. It has now become an extremely popular product. It is used in as diverse settings as maximum security prisons, national defense facilities, and even school bus parking areas. The systems are relatively easy and economical to install. Maintenance is minor. Malfunctions are rare. Repairs are fast and simple. The systems are excellent in discovering intruders with a minimum of false or nuisance alarms. They provide an excellent fringe benefit in facilitating layered systems, that is, to interface a variety of different sensors to achieve the same objective, thus providing synergistic success in intrusion detection.

The buildings within a facility constitute the second line of defense. The peripheral structures, (walls, ceilings, roofs and floors) constitute barriers. However, as is the case with perimeter barriers, they can be violated. Therefore internal intrusion detection systems are necessary to detect attempted or actual violations of the buildings.

A variety of popular systems are available to use alone or integrated into more comprehensive systems. For the sake of discussion, internal intrusion detection systems can be grouped into the area of sensing. The perimeter structures, portals and other openings can be equipped with sensors to detect unauthorized entry. Volumetric motion detectors can be installed to determine the presence of an intruder within an interior space. Invisible barriers (trip wires, infrared, and laser beams) can be used to determine when intruders venture beyond a given area. Point sensors can be employed to detect intruders approaching a given object.

The third line of defense is provided in critical areas. These are relatively small areas, structures including safes, vaults and strong rooms, wherein the most critical or sought after items are contained. They are constructed of heavily resistant material and, as we saw in Part II, intruder-resistant material can be augmented by active barriers. Yet with sufficient incentive, time, skill, and equipment, these structures can be violated. Many vaults contain millions of dollars, providing the intruder with an incentive. Time, skill and equipment, the remaining requisites, follow suit.

The last four articles in Part III provide a range of valuable and unique information on intrusion detection systems. The articles discuss either new systems, modified application of more commonly used systems, or improvement of existing systems. Although some of the articles contain formuli and equations, there is sufficient qualitative explanation to render them understandable to non-engineers.

The third article "A Ported Coaxial Cable Sensor for Interior Applications," is a totally new system for interior intrusion detection. The system, as we see in the first two articles, is well established for outdoor application; however, this is the first interior application. Yet its excellent detection qualities, coupled with low false/nuisance alarm rates, are even improved in interior application. It provides a combination of detection for peripheral structures, volumetric areas, point and critical areas. Obviously, it is extremely flexible and integrates well with other systems. This article explicitly presents application principles, test results, and the system's unique advantages.

Photo-electric beams have a long and successful history, described in the fourth article. With recent improvements and creative imagination, their potential broadens greatly. Photo-electric beams can be used in outside applications; most of their inherent drawbacks outside are overcome in inside applications. They integrate well with other systems, especially with ported coaxial cable. One of their unique potentials is they can be used as wireless communication links, which is also discussed in the article.

The fifth article, "New Developments in Ultrasonic and Infrared Motion Detectors," shows how these popular systems can be greatly improved with signal processing. Volumetric motion detection systems are economical and flexible; however, many professionals are concerned with their detection accuracy. They are especially vulnerable to false/nuisance alarms. Signal processing is becoming the state of the art to remove these weaknesses.

The sixth article is aimed exclusively at intrusion detection of critical areas. It discusses the application of piezoelectric microphones as seismic sensors. It specifically addresses protection of vaults and the unique problems of protection of automated teller machines. The tools and skill of intruders are becoming much more sophisticated. The challenge to security administrators is therefore becoming greater: seismic sensors may restore the competitive edge.

Burial Medium Effects on Leaky Coaxial Cable Sensors

R. Keith Harman

Abstract. A number of leaky (ported) coaxial cable sensors have been developed in recent years for use as buried line perimeter intrusion detection sensors. The performance of these sensors is affected by the electrical properties of the burial medium. While performance has been very good under a wide variety of installation conditions, a number of concerns about performance have been raised with regard to the effect of environmental conditions on the burial medium. For example, what effect does frost, snow and rain have on the sensor false alarm rate, nuisance alarm rate, probability of detection and range of containment? This paper describes the electromagnetic factors as they relate to sensor performance.

INTRODUCTION

While the electromagnetic principles described in this paper pertain to most leaky (ported) coaxial cable sensors, the research and experimentation relate directly to the SENTRAX product produced by Senstar Corporation. The background to this product was presented in a 1982 Carnahan paper entitled "Advancements in Leaky Cable Technology for Intrusion Detection."[1] A more detailed description of the product was presented in a 1983 Carnahan paper entitled "SENTRAX—A Perimeter Security System."[2] In order to make this paper self-contained, a brief introduction to the technology and the product is provided.

LEAKY CABLE SENSOR TECHNOLOGY

As illustrated in Figure 1, a basic leaky coaxial cable sensor comprises two buried parallel leaky coaxial cables. One cable is connected to a transmitter and the other is

1983 International Carnahan Conference on Security Technology, Zurich, Switzerland, October 4–6, 1983.

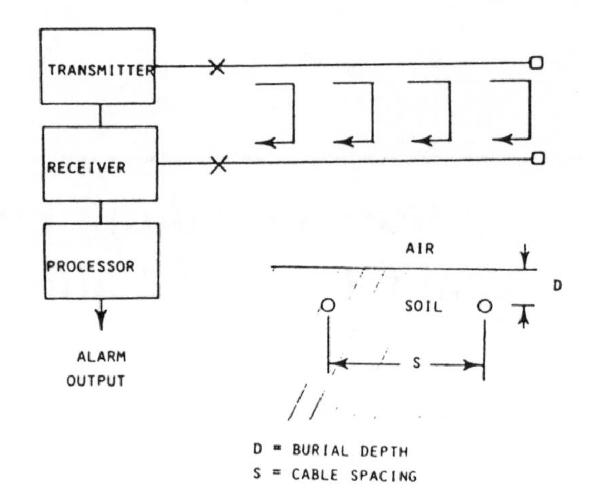

Figure 1. Basic leaky cable sensor.

connected to a receiver. The transmitter launches an RF wave down the transmit coaxial cable, thereby setting an electromagnetic field above the soil surface. An intruder in the vicinity of the buried cables causes another surface wave to propagate back towards the receiver. The reflected surface wave causes a field to be set up in the receive cable and hence at the receiver. The received signal is processed and a target is declared if the response meets the appropriate criteria.

In order to understand how a buried leaky coaxial cable sensor operates, it is useful to examine the various modes of propagation along such a cable. The signal

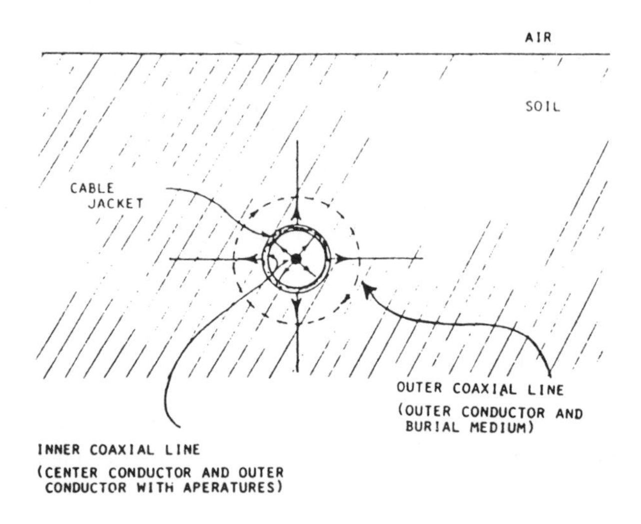

Figure 2. Modes of propagation along a leaky coaxial cable.

travelling inside the cable propagates in a normal coaxial mode with radial electric field lines and circular magnetic lines between the inner and outer conductor. There is an outer coaxial field produced by the energy which couples through the apertures in the outer conductor. Both these modes are illustrated in Figure 2.

While the inner coaxial cable is effectively a closed wave guide, in that it is set up between the inner and outer conductor, the outer coaxial field is supported by an open wave guide in that there is no outer conductor. In theory, the field lines extend to infinity, but the field strength decays with radial distance from the cable. The rate of this electromagnetic field decay is related to the permittivity and conductivity of the burial medium.

As described in a paper by A. S. Fernandes,[3] the outer coaxial cable mode can be represented by the following equations:

$$E_z = A_o H_o^1 (jh_o r) \quad \text{(longitudinal field)} \tag{1}$$

$$E_r = A_o \frac{\beta}{h_o} H_1^1(jh_o r) \quad \text{radial field)} \tag{2}$$

$$H_\theta = A_o \frac{k_o^2}{\omega \mu_o h_o} H_1^1(jh_o r) \quad \text{(coaxial)} \tag{3}$$

where r = radial distance
h_o, k_o, β = propagation factors
A_o = a constant
H_o^1 = Hankel function of first kind and order zero
H_1^1 = Hankel function of first kind and order one

The arguments of the Hankel functions are purely imaginary and can be replaced by Modified Bessel functions of the first kind and orders zero and one.

$$E_z = -jA_o \frac{2}{\pi} K_o(h_o r) \quad \text{(longitudinal field)} \tag{4}$$

$$E_r = -A_o \frac{2\beta}{\pi h_o} K_1(h_o r) \quad \text{(radial field)} \tag{5}$$

$$H_\theta = -A_o \frac{2\, k_o^2}{\pi \omega \mu_o h_o} K_1(h_o r) \quad \text{(coaxial)} \tag{6}$$

For small arguments

$$K_o(h_o r) \cong -\ln(h_o r) \tag{7}$$

and

$$K_1(h_o r) \cong \frac{1}{h_o r} \tag{8}$$

and for large arguments

$$K_o(h_o r) = K_1(h_o r) = \sqrt{\frac{\eta}{h_o r}}\, e^{-h_o r} \tag{9}$$

The radial component of the electric field decays with radial distance, r, relative to the field at a distance, r_1, as

$$E_r(r,r_1) = 20 \log_{10} [K_1(h_o r)/K_1(h_o r_1)] \quad \text{(dbs)} \tag{10}$$

From equation (8) one can expect a radial decay of 20 dbs per decade for small values of r and from equation (9) the field decays extremely rapidly for large arguments due to the exponential term.

It is postulated that this outer coaxial field excites a surface wave on the boundary between the air and the burial medium. This surface wave is illustrated in cross sectional view in Figure 3. This wave decays exponentially with height above the burial medium and propagates with a forward tilt angle. The forward tilt angle and the exponent of the decay factor increase with the losses in the burial medium. Hence, one can expect the field to decay more rapidly and at an increased tilt angle for a heavy clay medium than for a dry sand medium. A surface wave is said to be tightly bound to the air/medium interface when the exponential decay factor has a large exponent and loosely bound if it has a small exponent.

The sensor height response and rejection of small animals depends primarily on the degree to which the surface wave is bound to the air/medium interface. When the surface wave is tightly bound, the height response is less and there is less rejection of small animals. For loosely bound surface waves, the converse is true. In practice adequate performance is achieved even in highly lossy clay burial mediums.

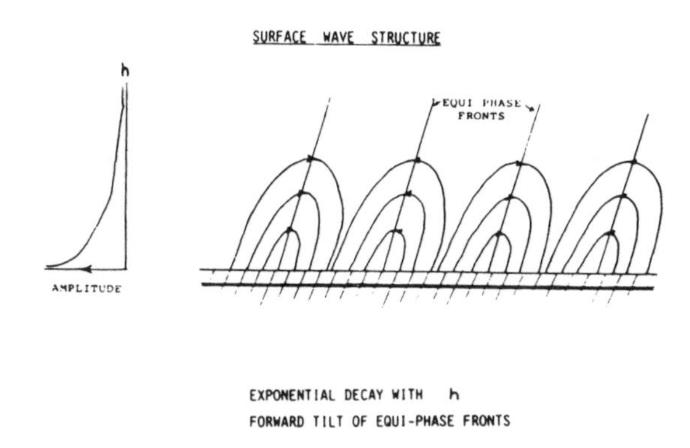

Figure 3. Surface wave on the boundary between the air and the burial medium, created by the outer coaxial field.

The range of containment of the sensor field also depends on the properties of the surface wave. This is due to the fact that surface wave structures form end-fire antennas. Radiation occurs from the start and end of the surface wave structure and at any discontinuity in the surface wave structure. The electromagnetic field tends to overshoot the corners due to this radiation effect. In other instances, discontinuity can cause the radiated field to extend beyond the desired detection zone. These undesirable effects can be controlled by following prescribed guidelines and procedures during installation of the cables.

PREDICTED AND EXPERIMENTAL RESULTS

The vertical component of the electric field was measured at ground level as a function of radial distance from the buried transmit cable. The results are plotted in Figure 4, along with a modified Bessel function approximation in which $r_1 = 2$ meters and h_o is assumed to be .270 radians per meter, up to a radial distance of 10 meters and inverse of radial distance beyond 10 meters. It would appear that a radiated field dominates beyond 10 meters while the coaxial mode and its related surface wave dominate up to 10 meters.

The extrapolated field decay function plotted in Figure 4 can be used to estimate the relative sensor response for a human intruder standing upright midway between buried cables. The sensor response will vary as a function of cable spacing. If a one meter cable spacing is used as reference spacing, the field strength at one half a meter on Figure 4 (17.5 dbs) becomes the reference field strength. The relative sensor response

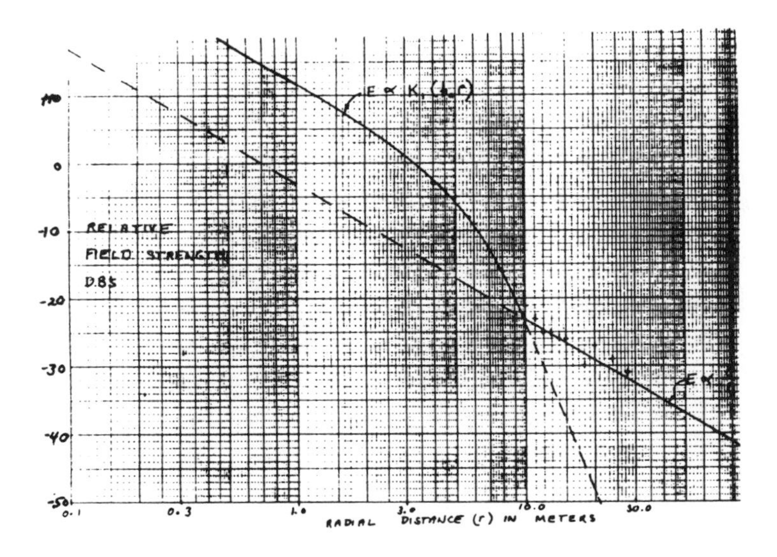

Figure 4. Extrapolated field decay function.

for a 2.1 meter cable spacing is then estimated by taking the relative field strength at 1.05 meters (11.5 dbs), subtracting the reference value (11.5 − 17.5 −6 dbs) and doubling the result to account for the return path (−12 dbs). Estimated responses for cable spacings of 2.1, 3.05 and 4.3, relative to the response for cables spaced at one meter, meters are −12, −19 and −25 dbs respectively. These compare favorably to the measured results of −11.5, −18 and −26 dbs.

The height response is also a function of cable spacing. In essence, the transmit field is most tightly bound to the air/medium interface directly over the transmit cable and becomes less bound with radial distance from the cable. The reciprocal condition applies for the receive cable and its susceptibility. Hence, one can conclude that the exponential decay factor of the surface wave structure for a target midway between the sensor cables will decrease with cable spacing to provide an increased detection height. Relative response to a human target at six different heights above the air/medium interface were measured and are plotted in Figure 5. The height response is based on a given threshold margin used during sensor calibration. For example, a margin of 12 dbs would provide height responses of 0.7 and 1.25 meters for cable spacings of 1.0 and 4.3 meters respectively.

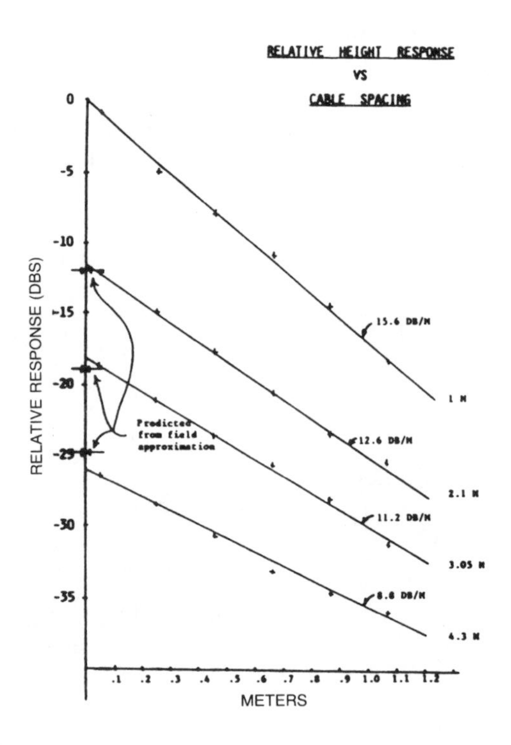

Figure 5. Relative response to a human target at six different heights above the air/medium interface.

SENTRAX PRODUCT DESCRIPTION

SENTRAX is a covert, buried line, terrain-following perimeter intrusion detection sensor. It has a secure redundant fail-safe power and data distribution network with color graphic video display and control unit. As a sensor, SENTRAX represents the second generation of electromagnetic buried line sensors utilizing leaky coaxial cables as the sensing device. It has been designed as a cost-effective answer to perimeter security for both large and small sites requiring varying levels of security under very demanding environmental conditions.

The SENTRAX sensor system is illustrated in Figure 6. It consists of six main components: buried parallel leaky coaxial Cable Sets (CS), Transceiver Modules (TMs), a Control Module (CM), an Operator Terminal (OT), a Printer (PR), and an Interface Unit (IU).

The TMs include the necessary electronic components to provide two zones of detection, and hence, are located between every other detection zone around the perimeter. The CM, which is located indoors, provides power to, and communicates with, the TMs using the sensor cables. The CM drives the OT and PR and receives input from the control keys to provide the all-important human interface.

The IU can be added to provide input and output ports to other sensors or devices.

SENTRAX TMs detect moving bodies or objects which have a high conductivity and adequate physical dimensions. This includes bodies of salt water and metal conductors which are longer than approximately one meter. Objects such as grass, trees, shrubs and wooden structures are not detected. The sensor threshold is normally adjusted to detect human intruders weighing more than 35 kilograms but not animals weighing less than 7 kilograms. The probability of detecting animals between 7 and 35 kilograms increases with mass.

Each TM detects and reports five different types of alarms: intrusion, tamper, cable fault, test failure and RF jamming. If a TM enclosure is opened, a tamper alarm is produced and if a cable is damaged, a cable fault alarm is produced. Each TM is equipped with a self-test capability and an associated test-failure alarm. The TM processor detects RF jamming and generates a unique alarm.

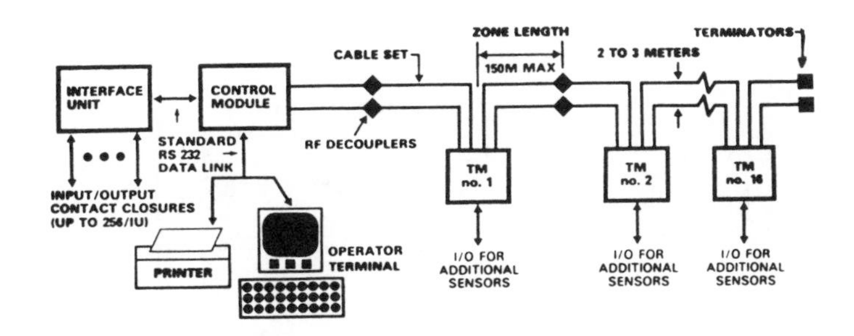

Figure 6. SENTRAX sensor system.

Power and Data Network

The power and data network formed by the buried cables is illustrated in Figure 7. SENTRAX (patent pending) is unique since it is used for intrusion detection, power distribution and data communication. Costly multi-conductor cables around the perimeter are not required to collect data. In a closed perimeter, the two parallel cables run in opposite directions around the perimeter from the CM. These cables can be viewed as two long "extension cords" around the perimeter which can both be cut without affecting the power and data network. The double redundancy feature is very important for a highly reliable system. Since the cables are also the sensor, this power and data network is secure as well as redundant and fail-safe.

In some cases, it is desirable to have more than one type of intrusion detection sensor on the perimeter of a layered system to achieve a high level of security. Other sensors can be connected to the SENTRAX TMs on the perimeter to receive power and communicate data. This extends the benefits of the secure, redundant and fail-safe power and data network to other sensors with the obvious cost savings.

The distributed data network can also be used to communicate with other devices on the perimeter. For example, it can be used to control gates and CCTV or to monitor weather conditions.

Control and Display

The CM collects the data from the perimeter sensor system and drives the OT and PR. In addition, the CM has a Liquid Crystal Display (LCD) for system diagnostics and as a backup to the OT. The microprocessor in the CM can be programmed to integrate the complete sensor system.

The OT color graphic display presents a map showing the layout of buildings and the site perimeter. The detection zones are numbered around the perimeter. When an intrusion occurs, a red rectangular box flashes at the appropriate location on the perimeter and an audible alarm sounds. The operator is prompted to acknowledge the

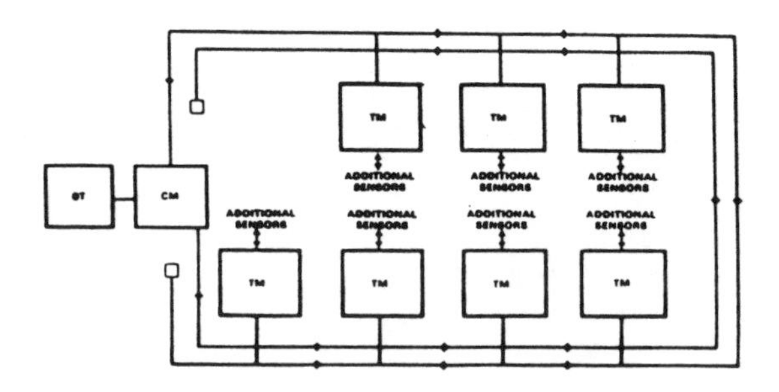

Figure 7. The power and data network formed by the buried cables.

alarm by pressing the designated key. Once the alarm is acknowledged, the red symbol stops flashing and the audible alarm is silenced. The operator is then prompted to reset the alarm or to access the zone depending on the situation.

A PR is connected to the CM to provide a data logging function. This records all alarms and operator actions for future use. An optional IU can be connected to the CM to accept data from, or supply data to, other subsystems such as fire alarms, interior sensor alarms and other perimeter sensors with centralized processors. Each CM can accommodate up to four IUs and each IU can provide up to 256 input or output points in multiples of 16, for a total of 1024.

Radio Regulations

Since leaky coaxial cable sensors produce electromagnetic fields, they must comply with the appropriate regulations. Due to the extremely narrow bandwidth of the SENTRAX continuous wave transmission, the U.S. FCC and the Canadian Department of Communications (DOC) have approved its use in the Industrial Scientific and Medical (ISM) band at 40.68 MHz. Since this band is internationally accepted, Senstar does not expect any difficulty in obtaining approval in other countries.

CONCLUSIONS

The design of the SENTRAX system is based on a sound appreciation of this relatively new leaky coaxial cable technology which was acquired through years of application experience and the results of fundamental experiments, such as those presented in this paper. As a second generation leaky cable sensor, SENTRAX provides many novel features which make it an ideal system integrator. Results to date indicate that SENTRAX provides excellent performance under a wide variety of environmental and installation conditions.

REFERENCES

1. R. K. Harman and J. E. Seidlarz, "Advancements in Leaky Cable Technology for Intrusion Detection," *Proceedings, 1982 Carnahan Conference,* University of Kentucky, Lexington, Kentucky.
2. R. K. Harman, "SENTRAX—A Perimeter Security System," *Proceedings, 1983 Carnahan Conference*, University of Kentucky, Lexington, Kentucky.
3. A. S. de C. Fernandes, "Propagation Charactaristics of a Loose Braid Coaxial Cable in Free Space," *The Radio and Electronic Engineer*, vol. 49, no. 5 (May, 1979), pp. 255–60.

Developments in Long-Line Ported Coaxial Intrusion Detection Sensors

Douglas J. Clarke
Michael S. Sims

Abstract. The first practical commercial ported coaxial cable intrusion detection system was installed for the Canadian Penitentiary Service and evaluated by Queen's University in 1977. An improved design was installed at a second site and independently evaluated in 1980. Both these systems employ pulsed rf technology in conjunction with a continuous long-line transducer cable.

Continuing research and development efforts since that time have led to considerable advances in system capabilities and installed performance for this unique design. During this period a number of other leaky cable intrusion systems have been developed, all of which employ continuous wave (cw) technology and thus use the earlier block sensor concept in site design.

Progress of these developments is reviewed, and some comparisons are drawn between the two approaches with respect to installed performance.

INTRODUCTION

The concept of ported coaxial cable intrusion detection sensors was first introduced in 1976.[1]* This was applied to a long-line prototype system, called GUIDAR, developed by Computing Devices Company (ComDev) for the Canadian Penitentiary Service (CPS) and evaluated by Queens University in 1977.[2] Subsequent improvements were made to the GUIDAR design and in 1979 a developmental system of this improved design was delivered to Correctional Services Canada (formerly CPS). This unit was

1984 Carnahan Conference on Security Technology, University of Kentucky, Lexington, Kentucky, May 16–18, 1984.

*Perimeter Surveillance System. Patent #4,091,367.

independently evaluated and the results were presented to the Carnahan Conference in 1980.[3]

A program, joint-funded between the United States Air Force Physical Security Systems Directorate (PSSD) and the Canadian Department of Industry Trade and Commerce, was awarded to ComDev (in 1978) for the Advanced Development of a long-line ported coaxial cable sensor (PCCS). This system was successfully evaluated against the U.S. tri-service requirements for an inventory item for application to external perimeter protection, and the decision to proceed with Engineering Development was made in 1982. This program is also joint-funded and will lead to a fully qualified military production item in the 1985 time frame.

Since these early developments a large number of GUIDAR installations have been successfully completed. Results from this large base of expertise have been fed back into a continuing research and development program at ComDev, aimed at both performance improvement and reduced costs. The first part of this paper provides an update in terms of current capabilities.

As the early GUIDAR developments evolved, a need was seen for a lower cost system for short perimeter applications. The result was a simplified cw design known as SPIR. The USAF undertook a comprehensive evaluation of SPIR for individual resource protection which was successfully completed in 1982. The SPIR system can also be applied to long perimeter applications by chaining them together for continuous coverage. However, this is not cost effective except for short segments, and has a number of other disadvantages when compared with the long-line sensor. The second part of the paper analyzes these trade-offs and discusses the benefits of the long-line sensor.

DESCRIPTION OF SYSTEMS

Scope

Short system descriptions of both GUIDAR and SPIR are included, together with some necessary additional information for SPIR that has not been previously published.

Common Characteristics

Both GUIDAR and SPIR employ ported coaxial cables (Figure 1) as the detection element. The cables are deployed along a perimeter a few feet apart, with the zone of detection defined by the cable routing. An rf field established between the cables is disturbed by the intruder. Processing electronics at the end of the cable detects the signal change and declares an alarm if the magnitude exceeds a predetermined level.

Since the electromagnetic field is contained relatively close to the pair of cables, the detection zone is truly terrain following. The major advantage of this technology is the generally superior performance relative to other sensor technologies, because of

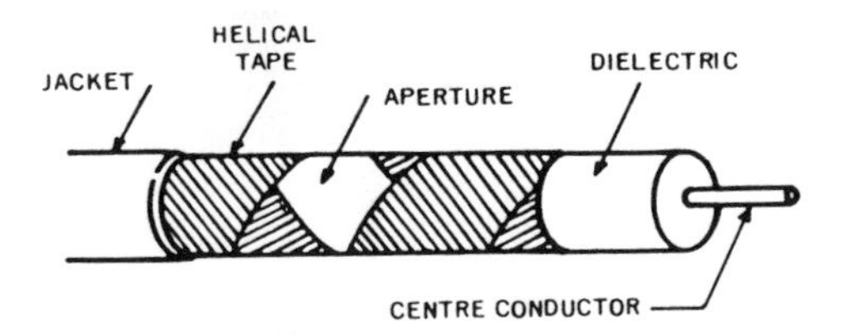

Figure 1. SPIRAX ported coaxial cable.

better immunity to environmental factors and rejection of small animals. Another key advantage is the ability to provide good detection regardless of topography. Many other sensors require considerable site preparation during installation and careful site maintenance during the system lifetime.

SPIR Characteristics

The SPIR system consists of an electronics unit which is capable of connecting to one or two pairs of Transducer Cables (Figure 2). Each cable pair can be up to 150 meters long, for a total length of 300 meters per system.

The unit operates from ac line input or 12V dc. Rf energy in the VHF band is alternately switched between the A side and B side transmit cables. The return on the receive cables is alternately processed through shared analog circuitry and digitizer with final digital processing performed by a 12-bit microprocessor. Detection thresholds and the resulting alarm outputs are provided for each side of the system. Since this system is intended for short perimeters, it operates in the switched cw mode and range information is not available. Consequently, location of the intruder is known only to be in the side of the system in which the alarm occurred.

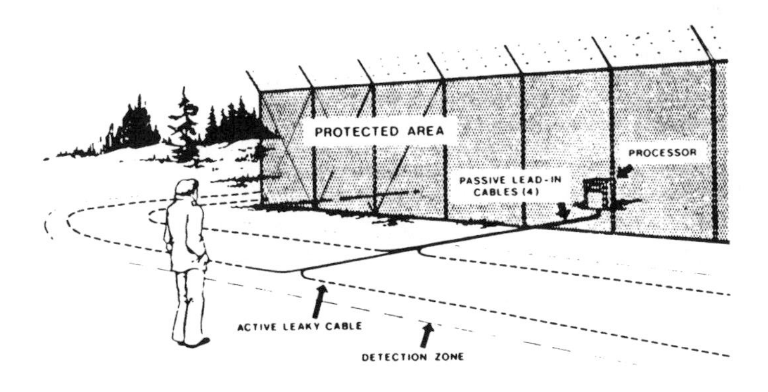

Figure 2. SPIR concept drawing for long perimeter application.

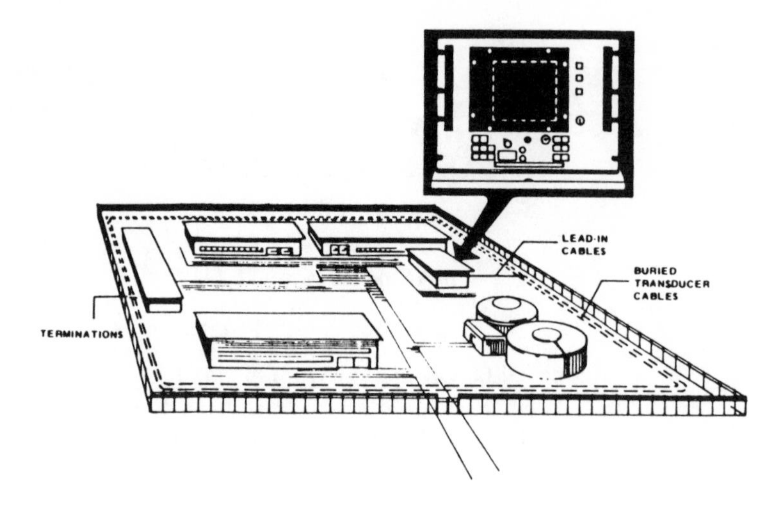

Figure 3. GUIDAR concept for 1600-meter system.

GUIDAR Characteristics

The GUIDAR system comprises a single processor unit deployed with one or two pairs of transducer cables (Figure 3), each of which can be up to 1600 meters in length. Hence a single processor is capable of covering perimeters up to 3200 meters (2 miles) in length.

GUIDAR employs pulsed rf transmission to determine intruder location along the length of the transducer cable by measuring the time delay of the return signal.

The perimeter is electronically divided into 33.3-meter range cells, each of which employs an individual detection threshold. These range cells are normally combined to provide alarm indication on a 100-meter zone basis.

PERFORMANCE IMPROVEMENTS

Methodology

A number of improvements have been introduced into the GUIDAR system since the first development model was evaluated in 1980. Paramount in this work was maintaining configuration control to ensure a design that would be easy to support in the years to come. Careful, original design has allowed significant improvements to be introduced over the past three years while maintaining full interchangeability and the ability to easily upgrade older systems as required.

The key to this is the use of plug-in modules for each major system function, designed to strict rules for interchangeability. Also, the use of microprocessors has allowed the majority of the implementation to be performed in firmware. Thus, major functions of filtering, signal processing and target detection can be significantly improved by merely updating the programmable-read-only-memory (PROM) devices.

Ported Coaxial Cable

One of the major changes to be introduced has been in the design of the ported coaxial cable. Since cable costs are a major part of the total system cost, new design techniques were sought to improve the method of cable manufacture. From very early in the evolution of GUIDAR the ability to model the cable and its interaction with the environment has been addressed. In recent times, the work carried out by Queen's University in conjunction with ComDev's IR&D efforts has resulted in the ability to accurately model the transducer cable. This work is described in a companion paper in this Proceedings.[4] The result of this work is ComDev's patented SPIRAX™ cable which is the first such cable to be designed for specific performance parameters, rather than developed by the cut-and-try experimental methods that had been previously employed.

It is essential to design a cable with the best possible characteristics because, regardless of the methods used in signal processing, the ultimate sensor performance can only be as good as the signals originating in the ported coaxial cables. While there are the obvious cost trade-offs to be made, modelling the cable performance as part of the design process ensures the best end result.

Processor Hardware

With the advent of Very Large Scale Integration (VLSI) in integrated circuit technology, considerable advances have been made in high-speed analog-to-digital conversion. One of the earliest changes was to update the Digitizer module to replace two large hybrid modules (which were state-of-the-art in the late seventies) with a single integrated circuit capable of 8-bit flash conversion at 15 MHz. The change was implemented while retaining the same form, fit, and function of the Digitizer module making it freely interchangeable with earlier models. Performance improvement obtained through this included considerably better linearity and superior reliability. The linearity improvement resulted in lower quantization noise and hence lower False Alarm Rate (FAR) for the same detection threshold setting.

Cable Monitoring

Since the transducer cable forms a critical part of the sensor, it is essential to monitor it continuously for correct performance and to detect attempts to tamper. While it is obvious that approaching the cables in an attempt to cut them would automatically result in detection, this in itself is not considered sufficient protection. The early GUIDAR systems employed a Line Terminator Unit (LTU) at the end of the transducer cable furthest from the Processor. This attenuated and delayed the transmitted pulse and fed the resulting signal into the receive cable. Hence a received pulse of the correct magnitude and time delay constituted a healthy system. However the technique requires this additional major component which must be installed at the end of the transducer cables.

From statistical data collected as a result of the early GUIDAR installations it became clear that the fixed signal return can be reliably used as a measure of system performance along the entire length of the perimeter. An algorithm was developed to perform this task and the LTU was replaced with simple resistor terminations.

Signal Processing

The main reason for the significant improvement in GUIDAR sensor performance over the past few years has been the application of high-speed digital processing to the detection function. This has made possible signal conditioning, automatic signature analysis and the use of spatial filtering.

For the GUIDAR system many tools have been created to develop signal processing techniques. These special research and development utilities have permitted raw data signals, occurring prior to GUIDAR processing, to be recorded on magnetic tape for subsequent processing and analysis. These tapes can be read directly by the ComDev in-house CDC CYBER 170/825 where the powerful features of this machine can be utilized to evaluate many different algorithms using real site data.

These utilities have already been employed for GUIDAR development, to generate new signal processing techniques in the areas of signal filtering, interference detection and target detection. The most significant contribution to performance and one that best demonstrates the potential capabilities of signal processing, is the Variable Zone feature.

Variable Zone Boundaries

To understand the purpose of this feature and its implementation, we must first describe the original method of target location. Since the GUIDAR Transducer Cable is a continuous length around the perimeter there are no physical sectors. For the purposes of target detection and subsequent location reporting, sectors are established electronically. This is done by taking digital samples of the return signal at a fixed time interval and, since the velocity of propagation of the cable is known, the samples relate directly to physical distance along the cable. This distance was chosen to be 33.3 meters and is known as a range cell. When three cells are combined for display purposes the system is analogous to a series of 100-meter block sensors.

From the considerable experience gained through installation of GUIDAR at many sites, the need for a flexible method of sector definition became apparent. This is particularly true of sites employing multiple sensors or closed circuit television (CCTV). We saw the opportunity to exploit the key benefit of the long-line sensor. Through a proprietary signal processing technique a method was developed to locate intruders with much better resolution. This permits sector boundaries to be located freely around the perimeter thus allowing variable length sectors. In addition, it is possible to have overlapping sectors or nested sectors whereby smaller sectors are contained within larger ones. It is anticipated that this feature will be used where it is desirable to call up multiple CCTV cameras in response to certain intrusion scenarios.

No physical modifications to the cables are required to implement the variable zone length feature. The feature is entirely a firmware/processing enhancement with no other hardware modifications required and, since no hardware has been changed, retrofit is possible for existing GUIDAR installations.

Alarm/mask zone definition and editing are accomplished via the existing GUIDAR front panel controls and displays. This provides on-site reconfiguration capability which has important life-cycle cost implications since the zone boundaries can be readily reconfigured as site conditions warrant with no physical change to the system. Six defined variable-length alarm zones can be designated as mask zones via the front panel. These zones can then be placed in access mode using the available switch actions on the front panel.

LONG-LINE SENSOR VERSUS BLOCK SENSORS

Introduction

Ported coaxial cable sensors have been developed to meet the users' need for a reliable intrusion detection device with low false/nuisance alarm rate that is economical to install and maintain. It is the intent of this section to examine two possible configurations that could satisfy the requirements for an average site. For each system concept, factors relating to installation, reliability, maintenance and performance will be examined to determine the relative merits.

The first configuration for review is the GUIDAR long-line ported coaxial sensor which has been designed specifically for long perimeter sites. This is compared with a second configuration using the SPIR short perimeter ported coaxial sensor. This can meet the needs of long perimeter sites by linking them together in a daisy-chain fashion. This is the traditional block sensor approach and, since the units are relatively low cost, would appear to be a viable solution.

Site Description

The average site assumed for this discussion has a perimeter just under one mile in length, defined by two parallel chain link fences. The perimeter is broken in at least one place for an entry point which utilizes a vehicle entrapment area. Additional security elements may include fence sensors and closed circuit television (CCTV). A primary display and control center with communications facilities is usually employed.

Although a large number of GUIDAR installations differ from the above (there is no average site) it is the opinion of the authors that using these assumptions provides a logical basis for discussion.

Typical GUIDAR Configuration

For the site as outlined the GUIDAR Processor unit would be located within the main control building. This unit operates with two pairs of ported coaxial cables known as A side and B side. For complete coverage of the perimeter the two pairs would start from the same point on the perimeter and travel in opposite directions to meet on the opposite side of the site (Figure 3). The two lead-in cable pairs would be installed below ground between the Processor and the nearest convenient point on the perimeter.

The key points to note are:

1. There are no externally located electronics.
2. Since the lead-in (non-ported) cable is an integral part of the Transducer Cable there are no externally located connections.
3. The alarm interface and system power are at a central point within the main building.

Typical Block Sensor Configuration

For the typical site whose perimeter length is 1600 meters (one mile) a minimum of six SPIR units would be required if we exploit the maximum capability of 300 meters per sensor. However, a more likely requirement is an average block length of 100 meters with a layout as shown in Figure 4.

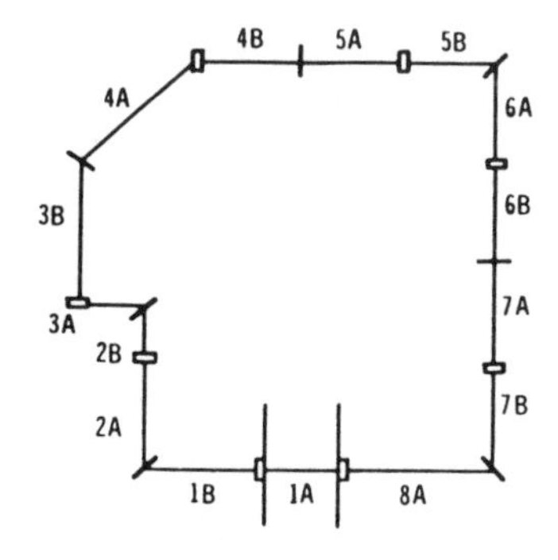

Figure 4. Layout of SPIR units on a typical 1600-meter site.

The 150-meter length provides for flexibility in trimming the zone to length. However, once installed, changing the sector lengths can be costly. This example will require eight units covering 15 blocks. Any special constraints (jogs, extra sectors, etc.) will increase the required quantity of units.

Installation Considerations

Both SPIR and GUIDAR have the same basic installation criteria for cable spacing and burial depth. In either case, two narrow parallel trenches can be cut all around the perimeter. However, in the case of SPIR, break-outs must be made to take the cable lead-ins out to each unit. For GUIDAR it is a simple matter to install the cables in one operation.

Additional effort required to install the SPIR system will include:

- Provision of a pedestal and poured concrete footings for the unit. Security considerations usually demand steel conduit for the cables.
- Individual connections for the four cables at each of the eight units.
- Individual calibration of each unit.
- Careful adjustment of the cable layout at start and end points.

The latter is necessary because the field developed in the two-wire mode takes some distance to build up. At the end point this mode can also travel a significant distance before it collapses sufficiently to prevent detection. Hence, care must also be taken at the junction between two units (including operation at different frequencies) to avoid detection by one unit of an intruder in an adjacent zone.

This adjustment is an important consideration since it impacts performance directly. Failure to carefully perform this operation will result in uneven sensitivity and high profile due to transmission line discontinuities. High profile can result in false alarms in changing environmental conditions.

Reliability Considerations

The main factors to consider in estimating the reliability of a system are the number of components, the quality level of components, the number of interconnections, the temperature range, and the shock and vibration.

A simple way to compare the number of components in the two systems is to consider the number of circuit cards. A GUIDAR Processor for this application with a Serial Interface will require fourteen cards. Each block sensor would require about two equivalent circuit cards per unit for a minimum of sixteen. This simple approach does not consider relative complexity. The biggest contributors to failure rate in component reliability are the microprocessor and associated memory devices. Eight block sensors will use eight microprocessors compared with two in GUIDAR.

High-quality military grade components are used in both GUIDAR and SPIR to ensure maximum reliability. However it is mandatory in SPIR to ensure that the unit will function over the temperature range that will be experienced in the outdoor environment.

Consideration of interconnections is particularly important for these sensors because of the microvolt levels that prevail within the rf sections. Even slight movement of contact surfaces will exhibit ohmic changes that are large compared with typical signal levels due to an intruder. Each such disturbance is one more false alarm that must be assessed. GUIDAR employs a connector at the end of each of the four Transducer Cables where they connect to the Processor, which is located in the benign indoor environment. SPIR, in this application, employs thirty-two such connections, all in the full outdoor environment.

The temperature in which the electronics operates plays a major part in the expected reliability. Calculating the reliability of GUIDAR at an ambient temperature of 25°C yields a Mean Time Between Failures (MTBF) of almost one year (24 hours per day, 7 days a week). This is a reasonable expectation since this unit will experience the temperature range in which the guard force is expected to work.

In calculating the MTBF for SPIR the full outdoor temperature range must be employed. Increasing the ambient temperature from 25 to 50°C doubles the component failure rate. Assuming that the total system component count for eight SPIR units is equivalent to one GUIDAR unit (very optimistic) and ignoring the connector reliability factor, we are now looking at a system MTBF of less than 6 months. This impacts maintenance, as we shall see.

Shock and vibration is not a reliability factor for these systems since the environment is ground, benign and, in any event, is the same for both.

Maintenance

Since both SPIR and GUIDAR have been designed not to require Preventative Maintenance (other than recalibration for major climatic changes), only repair action will be considered here.

Users have raised an important consideration when it comes to comparing the impact of failure of an entire GUIDAR system versus a single block in the SPIR case. This cannot be dismissed lightly, particularly when considering the operating procedures required as back-up during failure of electronic security. However, there are a number of factors that must be carefully examined before the final conclusions are drawn.

First consider the probability of failure and the Mean Time to Repair (MTTR) for these systems. For GUIDAR the Reliability calculation shows that it will not fail more often than once a year, and experience since 1980 with over twenty systems has shown that more than double this time is typical.

Repair action for GUIDAR is simple because of the centralized nature of the system and the powerful diagnostics available. The repair technician, in relative comfort, can quickly isolate the failed module and substitute a plug-in replacement within a few minutes. Alternately, if skill levels are limited, an entire Processor can be replaced in minutes.

Compare this with the unfortunate maintenance man responsible for a string of block sensors. Murphy's law ensures that the unit will fail on the darkest, coldest night of the year. In fact this is inevitable since the calculations show that there will be at least two failures per year and stresses are highest at extremes of temperature.

The procedure to effect the repair will be to walk out to the failed sector, probably violating security of the other sectors on the way, and exchange units under the vigilant eyes of the potential intruder. Any occurrence of a persistent problem will quickly be known as a vulnerable area of the perimeter.

Performance

Although SPIR and GUIDAR employ the same basic technology and exhibit the same general performance, there is considerable potential for superior performance with GUIDAR in the long perimeter role.

The first factor, already discussed, is the probability of sensitivity variations and abnormal profile due to disruption of the two-wire mode at the junction of each block for the SPIR system. This is compounded by the limitation of the cw system in terms of detection thresholding. That is, there can be only one threshold value for each block regardless of length. By comparison, GUIDAR establishes thresholds on a cell basis which is currently every 33 meters. The shorter threshold segment also enhances performance where media-induced sensitivity variations occur. Where these changes are large, an animal in a high-sensitivity region can appear to be as large as a human intruder in a low-sensitivity area. Obviously, the shorter the length of the threshold segment the better the performance is likely to be.

Another factor fundamental in the difference between pulsed and cw systems is the ability to establish sector boundaries electronically and independently of physical site conditions. This considerably simplifies initial installation since the required boundaries are easily set after construction is complete. During the life of the system, boundaries can be easily reconfigured for subsequent site changes or alignment with new or modified complementary sensors and cameras.

Vulnerability of a block system should also be considered. While it may be possible to bury each unit in an underground vault, cost and accessability during winter normally preclude this. The location of the unit must be carefully considered since, even if it is within its own protected area, it is visible and subject to sabotage and tamper.

Although all the preceding arguments are probably sufficient to demonstrate the advantages of a long-line sensor when compared with a block-type sensor, the real bonus comes from the potential derived from signal processing based on data available from the entire system.

The ability to analyze signal characteristics to discriminate between animals, humans, environmentally induced changes and other man-made external sources presents enormous potential for performance improvement. A pulsed long-line system is unique in its ability to perform this type of signal processing because the essential ingredients are there, namely, the ability for one processor to "see" the complete perimeter and thus be able to perform both spatial and time domain analysis.

By comparing data in each cell (33 meters) along the length of the Transducer Cable, it is relatively simple to see differences between an intruder crossing in one location compared with an environmental change occurring across many cells. If phase angle as well as magnitude is included, some startling differences between various signal sources become apparent. Performing this type of processing prior to declaring an alarm yields a higher Probability of Detection together with a lower nuisance/false alarm rate.

Summary

In the second part of this paper the long-line ported coaxial sensor has been compared with a string of block sensors around an average 1600-meter site. This has been summarized in Figure 5, where it can be seen that GUIDAR is superior in all cases.

FACTOR	GUIDAR	BLOCK SENSOR
External electronics	- None	every 200 metres
Installation		
- external mounting hardware	- None	Pedestal and conduit for each unit
- cable adjustment	- start & end of system	start & end of each zone
Reliability		
- electronics	7000 hours	3500 hours
- connectors	4 interior	32 exterior
Maintainability	card exchange in main building	unit exchange on perimeter
Performance		
- threshold segment	33 metres	100 metres
- spatial filtering	YES	NO
- match filtering	YES	NO
Detection Zone		
- range of lengths	5 - 1600 metres	50-150 metres
- modifiable after installation	YES	NO
- overlap or nesting	YES	NO

Figure 5. Comparison summary for 1600-meter perimeter.

The Last Word

Since SPIR has been compared with GUIDAR in the long perimeter role and found to have inferior characteristics, it must be emphasized that it is only in this context that this is so.

SPIR was designed for individual resource protection or short segments of long perimeters. For these applications it is eminently suited. It is an extremely cost effective solution for a reliable stand-alone sensor with all the favorable performance attributes for which ported coax is known.

CONCLUSIONS

In the first part of this paper some of the performance improvements were described that have been incorporated into GUIDAR over the previous three years. This showed that, not only are some significant benefits already available through signal processing in long-line sensors, but that there is a large potential in the future for further progress in this direction.

In the second part of this paper a comparison was drawn between GUIDAR, a long-line ported coaxial sensor, and SPIR configured in a multiple-sensor block system, for long perimeter applications. In its current configuration, GUIDAR has been shown to be vastly superior in terms of performance, reliability, maintainability and ease of installation. Furthermore, the fundamental design approach provides the opportunity for significant developments in signal processing for improved performance.

REFERENCES

1. R.K. Harman and N.A.M. Mackay, "GUIDAR: An Intrusion Detection System for Perimeter Protection," *Proceedings, 1976 Carnahan Conference on Crime Countermeasures.*
2. *GUIDAR Evaluation: A Line Sensor for Perimeter Surveillance*, by Queen's University (Kingston, Ontario), 30 Sept 1977.
3. R.D. Ball, "Evaluation of the GUIDAR Intrusion Detection System for the Correctional Services Canada (CSC)," *Proceedings, 1980 Carnahan Conference on Crime Countermeasures.*
4. M.C. Maki, "Coupling Mechanisms in Leaky Cable Sensors," *1984 Carnahan Conference Proceedings.*

A Ported Coaxial Cable Sensor for Interior Applications

Ronald W. Clifton
Rexford G. Booth

Abstract. A new interior sensor that utilizes ported coaxial cable technology is currently under development by Computing Devices Company for the U.S. Army Belvoir Research and Development Center. The development program is jointly funded by the U.S. Army and the Canadian Government's Department of Regional Industrial Expansion. This new sensor has been designated the Ported-Coax Interior Sensor (PINTS).

Ported coaxial cable sensors that have been developed to date for Department of Defense applications have been exterior perimeter surveillance sensors. These sensors have been developed to provide all-weather performance in the very demanding, and often hostile, outdoor environment.

The interior/indoor environment is significantly more benign, offering major advantages in sensor configuration and the selection of detection criteria. In addition, the ported coax technology has demonstrated immunity to the major false/nuisance alarm sources common to other interior sensors. The major challenge in developing a ported coax sensor for interior applications is to accommodate the wide range of operational and installation parameters that will be encountered during deployment.

This paper will introduce the technology as it applies to the interior environment. The operational concept for the sensor will be discussed and the results of preliminary testing will be presented.

INTRODUCTION

The Department of Defense (DOD) has a stated operational need for an effective physical security system for interior areas of facilities. Such a system is needed to prevent unauthorized forcible or surreptitious access to protected areas, removal of resource items and/or access to sensitive and classified areas.

1985 Carnahan Conference on Security Technology, University of Kentucky, Lexington, Kentucky, May 15–17, 1985.

The U.S. Army Belvoir Research and Development (R&D) Center located at Fort Belvoir, Virginia, has the responsibility of coordinating the long-range tri-service efforts to develop such a system. An early system, designated the Joint-Services Interior Intrusion Detection System (J-SIIDS) has been developed for the protection of arms rooms and is currently in use.[1] The Facility Intrusion Detection System (FIDS) has also been developed by the Belvoir R&D Center and is currently under operational test. The FIDS comprises a centralized command, control and communications console, a standardized interface protocol and a variety of standardized interior sensors.[2,3]

The Ported-Coax Interior Sensor (PINTS) is currently under development by the Belvoir R&D Center as an advanced FIDS sensor. A multi-phase development program has been defined and is jointly funded by the U.S. Army and the Canadian Government's Department of Regional Industrial Expansion (DRIE) under the U.S./Canada Defense Development Sharing Agreement. The contractor is Computing Devices Company, a Division of Control Data Canada, Ltd.

The PINTS has successfully completed Phase I. Feasibility Test and Evaluation. The sensor is currently in Phase II: Advanced Development.

TECHNOLOGY BACKGROUND

Ported coaxial (''leaky'') cables were originally developed for use in distributed communication systems. In the early 1970s, Computing Devices Company started applying this technology in a novel way to the detection of vehicles and humans. This work has lead to the development of sensors for a variety of high-value physical security applications.[4,5,6]

Similar in construction to standard coaxial cables, ported coaxial cables have holes or apertures in the outer shield spaced at regular intervals. These apertures allow a controlled amount of rf energy to couple in and out of the cable. Hence, a ported coaxial cable can be thought of as a coupled transmission line that can be used as either a transmitting or receiving transducer element. Design considerations for ported coaxial cables used in security sensors have been described in a previous paper.[5]

To date, ported coax sensors have been developed exclusively for outdoor perimeter surveillance applications. The major benefits of this technology have been the increased Probability of Detection (P_D) and reduced False/Nuisance Alarm Rate (FAR/NAR) obtainable in the very demanding and hostile all-weather environment. The PINTS is the first application of this technology to interior applications.

The interior application of the ported coax technology represents a significantly more benign operating environment that results in a major improvement in P_D and FAR/NAR performance over state-of-the-art outdoor sensor performance. In addition, relaxation of environmentally-driven design parameters allows more flexibility in sensor configuration and the selection of detection criteria.

Although the indoor environment offers improved sensor performance, there is an additional challenge to be addressed. The interior sensor application represents a wide range of operating and installation parameters:

- Size, shape and construction of the secure area;
- Threat definition;
- Type, value and nature of resource items.

Hence, the development of a ported coax sensor for interior applications must adequately accommodate this wide range of operating and installation conditions.

A ported coax sensor comprises one or more pairs of transmit/receive transducer cables and associated signal processing electronics. For the PINTS interior application, the cables will typically be mounted on a surface in the secure area. The location/placement of the cables in the secure area defines the detection zone for the sensor.

A generalized cross-sectional representation of a cable pair installation and the major parameters that affect detection zone size are provided in Figure 1. The major installation parameters are the cable-to-cable spacing (D), the cable-to-mounting-surface distance (S) and the electrical properties of the surrounding media. Not shown, but also important, are the cable length and the method of termination. The interrelationships between these major factors determine the longitudinal and cross-sectional detection zone(s) for the sensor.

For a given mounting surface type, the optimum cable-to-cable spacing is a function of both operating frequency and transmitter power. The cable-to-mounting-surface distance is very dependent on the electrical properties of the mounting surface as well as operational considerations (e.g., accessibility). In defining the PINTS detection zone it is important to view the transducer cables and the media as a coupled

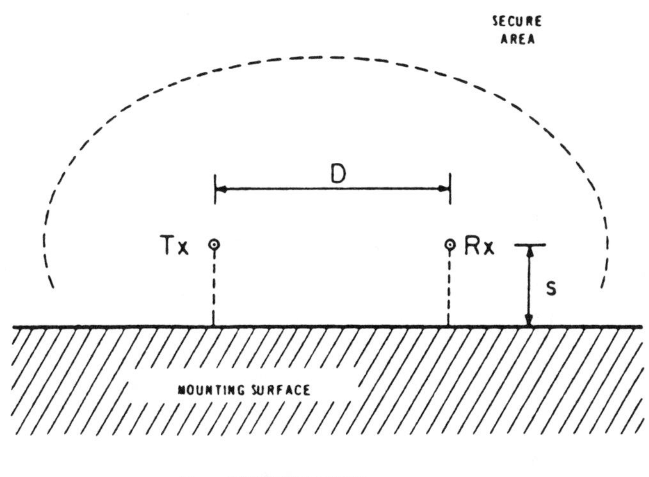

Tx = TRANSMIT CABLE
Rx = RECEIVE CABLE
D = CABLE-TO-CABLE SEPARATION
S = CABLE-TO-MOUNTING SURFACE SEPARATION

Figure 1. Generalized cross-sectional representation of transducer cables.

transmission line. Therefore, consideration of the surrounding media is essential for both the design and installation of this sensor.

In the construction of secure facilities for DOD applications, specific materials are required depending on the category of resource being stored.[7] Concrete and reinforced concrete structures are common. Other construction materials such as sheet metal or wood are also used, but are less common.

The range of construction materials along with other operational considerations (e.g., ease of installation) presents the security systems planner with a variety of options for installing the transducer cables. For many materials (e.g., reinforced concrete, concrete and wood) the cables can be mounted directly on the surface using standard tools and fastening devices. For covert operation or to make the cables inaccessible to traffic, the cables can be embedded in the material (e.g., concrete, reinforced concrete) or, in the case of rf transparent materials, behind it. For some installations (e.g., sheet metal) the cables must be mounted away from the surface for proper operation. Examples of these installation configuration options are shown in Figure 2. In studying these examples it should be noted that, although only two cables are shown for clarity, the techniques illustrated apply equally well to more than two cables. Furthermore, more than one configuration may apply for a particular facility.

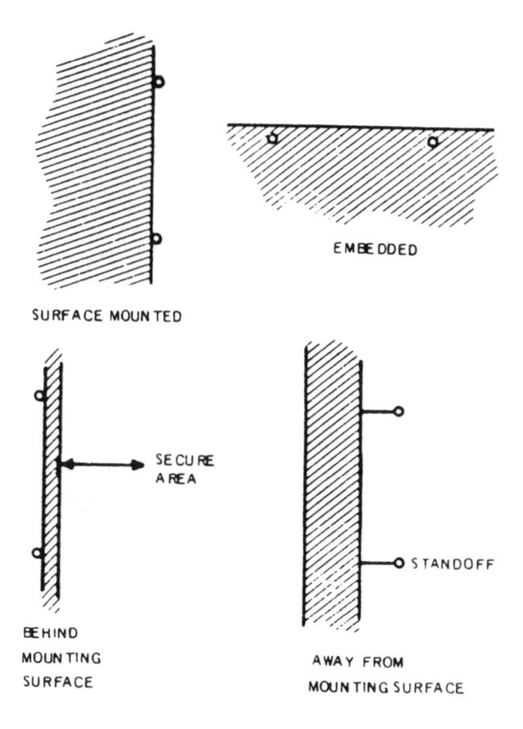

Figure 2. Optional installation configurations for transmit/receive cables (media dependent).

OPERATIONAL CONCEPT

A typical DOD facility may have a very large number of interior/secure areas to be protected. To reduce manpower requirements and increase vigilance, intrusion detection sensors are deployed in each secure area. For a large number of areas, however, even a moderate FAR/NAR for each sensor can be amplified into an unacceptably high false dispatch rate for the apprehension/response forces. When the false dispatch rate becomes excessive, the effectiveness of the total security system becomes jeopardized.

Therefore, there is an operational need for a reliable, high-performance intrusion detection sensor. Such a sensor must offer flexibility of configuration to meet a variety of installation requirements. The sensor must provide a high P_D and very low FAR/NAR in order to be effective.

A block diagram showing the major functional components of the FIDS and the PINTS interface to the integrated security system is provided in Figure 3. The FIDS operator(s) interacts with the security system through the Control Communication Display System (CCDS). Individual sensors in each remote/secure area are connected to the system through a local Control Unit which provides uninterrupted power, and also performs data collection and status monitoring functions. In addition to sensors,

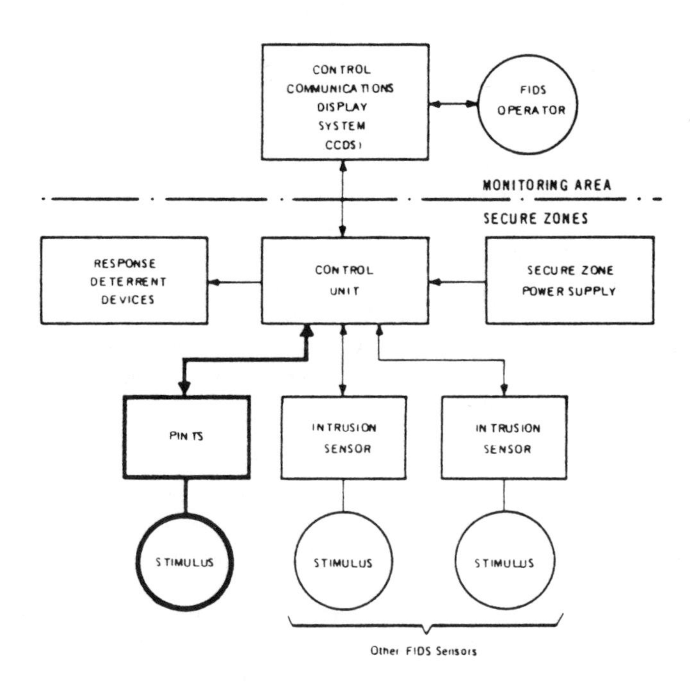

Figure 3. Major functional components of the Facility Intrusion Detection System (FIDS).

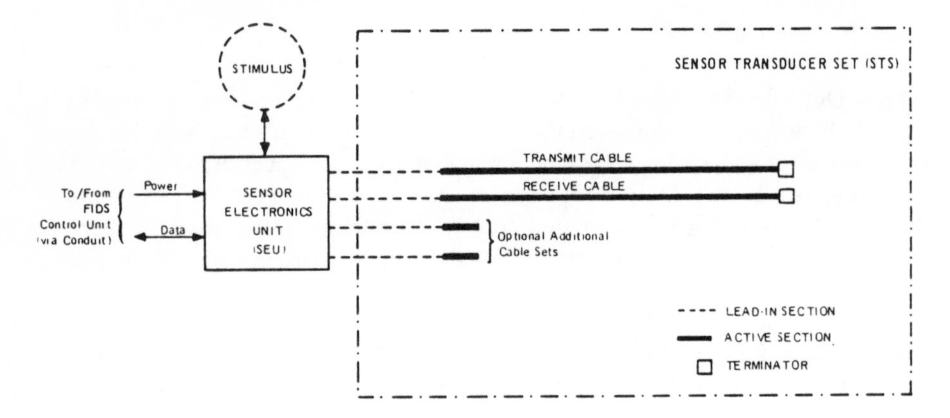

Figure 4. Major functional components of the PINTS.

response/deterrent devices may be installed and activated through the Control Unit. Each sensor also includes a stimulus which can be remotely activated from the CCDS to verify sensor operation.[2] Current FIDS sensors operate using a wide variety of detection principles (e.g., passive infrared, ultrasonic and vibration).

As shown in Figure 4, the PINTS comprises the following major functional components:

Sensor Electronics Unit. The SEU is a stand-alone unit which contains the necessary transmitter, receiver and processing electronics to interface with the transducer cables and perform the detection of threat-level targets. The SEU obtains prime power from the FIDS and uses the same interface to receive commands/ controls and to report alarms/status data. The SEU can be located either inside or outside the protected area, depending on convenience and base security procedures. A line drawing of the Advanced Development Model of the SEU is provided in Figure 5.

Stimulus. The PINTS Stimulus is contained in the SEU. Upon command from the FIDS, the Stimulus is activated to simulate a threat-level target signature. The Stimulus effectively tests the SEU, the transmit cables and the receive cables. The Stimulus can be activated independently for one or all installed cable sets.

Sensor Transducer Set. The STS comprises one or more (maximum of 4) independent pairs of ported coax cables. Each cable pair includes one transmit and one receive transducer cable complete with integral non-leaky lead-in cable. The receive and transmit cables are identical and, as such, are completely interchangeable. The cables are small diameter, designed specifically for the interior appli-

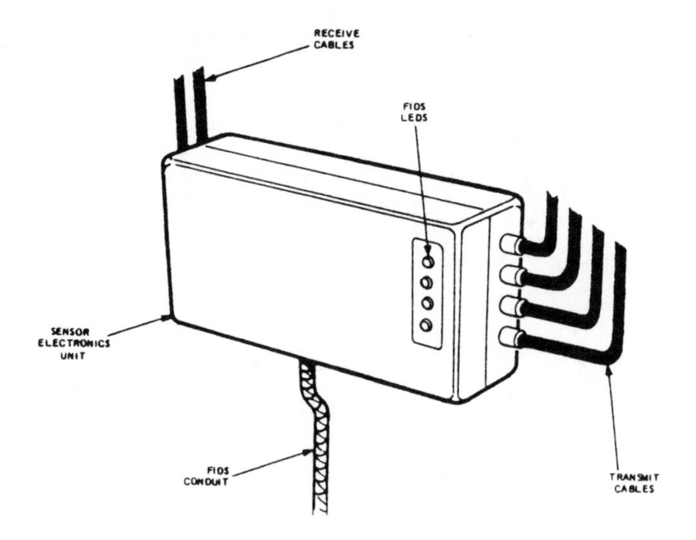

Figure 5. Outline drawing of the PINTS sensor electronics unit (advanced development model).

cation.[5] The cable sets can be cut to length so as to tailor the detection zone to the desired configuration.

One of the major features of the PINTS is its configuration flexibility. With up to four independent cable pairs, the PINTS can be configured to provide a variety of optional detection zone geometrics. The conceptual drawings of Figures 6, 7 and 8 are provided to illustrate the wide range of applications for which the PINTS is intended.

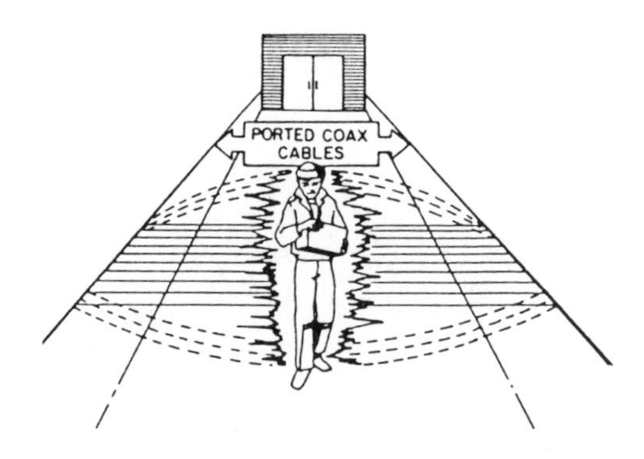

Figure 6. PINTS hallway/tunnel concept.

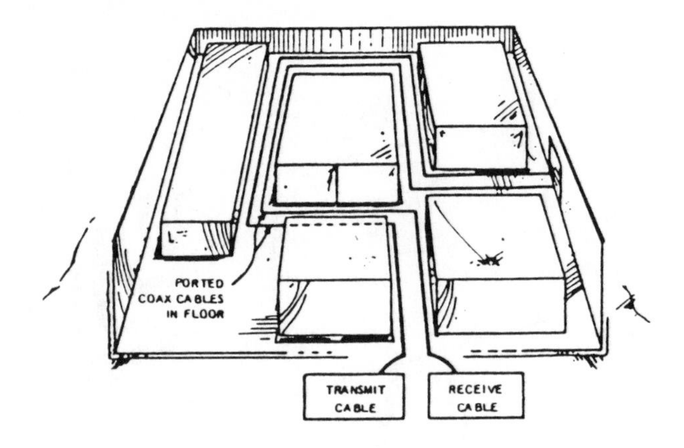

Figure 7. PINTS warehouse application concept.

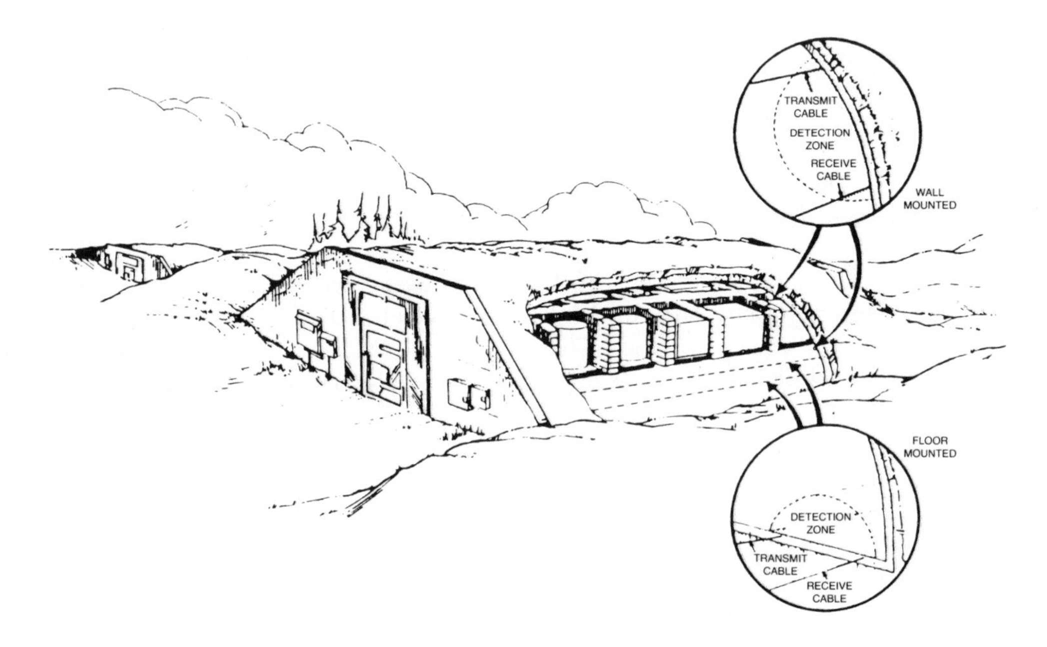

Figure 8. PINTS storage magazine concept.

OPERATIONAL BENEFITS

A major operational feature of PINTS is that the detection zone is effectively confined to a region in the immediate vicinity of the ported coax transducer cables. Since the detection zone is defined by the spacing, location and length of the transducer cables, the PINTS can be thought of as a user-configurable ''guided radar'' sensor. Major benefits of guided radar operation include the following:

Guided Detection Zone. Since the PINTS detection zone is guided, the sensor is directly applicable to irregularly shaped areas and avenues of approach. The shadow effects common to line of sight sensors (e.g., microwave, ultrasonic, infrared) are therefore eliminated. The PINTS is also well-suited to long-line applications such as tunnels, hallways, stairwells and maintenance shafts where geometry and/or corners may preclude surveillance by other types of sensors. The sensor can also be configured to provide ''perimeter'' surveillance inside large facilities.

Confined Detection Zone. A well-contained detection zone is effectively established by placement of the ported coax cables. This feature ensures that nuisance alarm traffic outside the zone of interest can be selectively excluded (e.g., active areas).

User Adaptable. The designer of an overall security system has considerable flexibility in configuring the PINTS. Major parameters under control of the designer are: location of the electronics unit; length of lead-in (non-active) cables; the length and spacing of the ported coax transducer cables; and, the number of independent cable pairs. In essence, the PINTS can be trimmed to fit the application.

Another major feature of the PINTS is its VHF operation. Practical use of ported coaxial cables for intrusion detection is limited to the low VHF range. VHF operation provides the following additional operational benefits:

Optimum Detection of Humans. A typical threat-level human is a significantly large fraction of a wavelength at these frequencies. This provides optimum detection of humans while rejecting small animals.

Rejection of Common FAR/NAR Sources. VHF operation ensures that the PINTS is not susceptible to false/nuisance alarms caused by acoustic noise, air motion, dust or debris. In addition, by virtue of its relatively long wavelength, the sensor is not adversely affected by seismic or other vibration effects.

Complementary to Other Technologies. The PINTS, as a VHF motion sensor, is complementary to other intrusion detection sensor technologies (e.g., microwave, vibration, ultrasonic, infrared). Hence, it is intended that the PINTS can be collocated with other existing interior sensors. Collocation of sensors offers operational performance advantages for applications with overlapping detection coverage and other integrated sensor requirements.

These combined features provide PINTS with performance advantages, flexibility of configuration and other operational benefits not currently available from any other interior sensor.

FEASIBILITY TEST AND EVALUATION

The major objective of the PINTS Phase I program was to perform sufficient analyses and testing to determine the feasibility of this new sensor. Much of the fundamental work was done using Computing Devices Company facilities. Major feasibility testing was conducted at designated Fort Belvoir test sites. Over the period of the Phase I program, a total of 8 weeks of on-site test and evaluation were successfully completed at Ft. Belvoir.

The Fort Belvoir facilities comprised buildings A, B, C, D and E as follows:

- *A*—60' reinforced concrete magazine;
- *B*—40' reinforced concrete magazine;
- *C*—sheet metal ''Butler'' building;
- *D*—wooden storage building;
- *E*—brick/concrete office building.

These facilities were selected because they were either operationally representative or provided good test conditions. Building C, for instance, is known to cause a high FAR/NAR for other sensors under moderate to heavy wind conditions. These five facilities also presented a wide range of building construction materials representative of ''real-world'' applications.

Simulated operational testing was conducted to determine the performance capabilities of the sensor. A total of 14 sensor installations/configurations were tested. Evaluation of each installation/configuration comprised the following:

> P_D *Tests*. These tests were intended to simulate a variety of threat scenarios. Major target/intruder parameters varied were: facility access point, intrusion path and length, intruder orientation (upright, crawl, roll), intruder speed and intruder mass.
>
> *FAR Tests*. These tests were performed for each configuration and facility. Tests consisted of monitoring the sensor FAR response for extended periods of time. All FAR tests included full assessment by ''stump-sitters'' and periodic intrusions to verify operation.
>
> *NAR Tests*. Due to time constraints limiting FAR testing, NAR tests were performed to effect accelerated testing of FAR/NAR performance. NAR tests included heavy external vibration, small animals, RF/electromagnetic interference (RFI/EMI) and external movement (vehicles, humans and animals).

All simulated operational tests were witnessed by the U.S. Army and were conducted in accordance with an approved test plan.

All sensor detection parameters were determined for each configuration during initial installation and then kept constant for all subsequent testing. Within the limitations of the test equipment used, sensor response for each P_D and FAR/NAR test was monitored to provide a quantitative measure of performance. In addition to alarm on/off information, typical parameters measured included target response relative to threshold and target Signal-to-Noise-Ratio (SNR).

A summary of the test results is as follows:

> *P_D Summary*. A total of 409 intrusions were attempted. For many configurations all intrusions were detected, resulting in 100% P_D when calculated as a simple percentage. Statistically, based on the total number of tests performed, P_D ranged from 94 to 99% with a typical confidence level of 95%. A typical P_D test summary is provided in Figure 9.
>
> *FAR Summary*. No false alarms were observed in 94 hours of FAR testing. The testing periods were continuous up to 24 hours and included periods of heavy thunderstorm activity.
>
> *NAR Summary*. A total of 59 NAR tests were performed. Of these, only one test caused any significant response. This anomaly was observed for one configuration in one facility only and is currently being addressed. Overall, the NAR tests demonstrated excellent rejection of vibration and wind effects, RFI/EMI, small animals and external activity.

Based on these test results and a study of operational considerations, it was determined that the PINTS concept is feasible and practical.[8] The major recommendation that resulted from the Phase I study was that the PINTS should proceed directly to Phase II (Advanced Development).

Test	Site	Detec-tions	Misses	Average SNR (dB)	
				Left Pair	Right Pair
Normal Speed Intrusions	B	30	0	43	47
Crawling Test	B	15	0	32	40
Stay-Behind/ Wall Intrusion Test	B	20	0	49	44
Normal Speed Intrusions	A	12	0	37	33
Running Test	A	12	0	34	30
Slow Walk Test					
- 12 inches per second	A	6	0	37	33
- 3 inches per second	A	6	0	34	31
Crawling Test	A	12	0	28	26
Stay-Behind Wall Intrusion Test	A	30	0	33	35
Intrusions with Storage	A	12	0	28	36

SNR - Signal to Noise Ratio
dB - decibels

Figure 9. Typical P_D test summary sheet.

CONCLUSIONS

This paper has been an attempt to summarize the development and testing to date of a new interior intrusion detection sensor being developed for tri-service applications. The material presented demonstrates that the PINTS offers the potential of performance and operational benefits currently not available from any other interior sensor. It is expected that the PINTS will significantly contribute to the repertoire of standardized FIDS sensors.

REFERENCES

1. R.L. Barnard, ''Application for the Joint-Services Interior Intrusion Detection System,'' *Proceedings of the 1974 Carnahan and International Crime Countermeasures Conference*, April 1974, pp. 53–8.
2. B. Barker, and R.A. Miller, ''Facility Intrusion Detection Systems,'' *Proceedings of the 1980 Carnahan Conference on Crime Countermeasures*, May 1980, pp. 163–70.
3. ''Interior Facility Intrusion Detection System (FIDS),'' Combined Arms Support Laboratory, STREBE-XI, Fort Belvoir, Virginia, March 1983.
4. D.J. Clarke, and M.S. Sims, ''Developments in Long-Line Ported Coaxial Intrusion Detection Sensors,'' *Proceedings of the 1984 Carnahan Conference on Security Technology*, May 1984, pp. 5–12.
5. M.C. Maki, ''Coupling Mechanisms in Leaky Cable Sensors,'' *Proceedings of the 1984 Carnahan Conference on Security Technology*, May 1984, pp. 195–201.
6. R.W. Clifton, and R.E. Patterson, ''Ported Coaxial Cable Configuration Study for a Remote Site Security Sensor (RSSS),'' Final Technical Report, RADC-TR-81-344, November 1981.
7. Department of Defense, ''Physical Security of Sensitive Conventional Arms, Ammunition and Explosives (AA&E),'' DOD 5100.76-M, March 1982.
8. A.R. Schmidt, D.T. Taylor, and R.W. Clifton, ''Ported-Coax Interior Sensor Feasibility Study (PINTS Phase I),'' A226/A007/FR, Ft. Belvoir, Virginia, November 1983.

Opportunities for Photoelectric Beams for Indoor and Outdoor Security Applications

David L. Andrew

Abstract. Photoelectric beams have been used in security applications for more than 50 years, and are still seeing growth possibilities, through many cost reductions, improvements and competition from other technologies.

This article looks at the future of active photoelectric beam use in the security industry through an historical examination of its development, its operational characteristics and its configurations. Some of the main advantages and drawbacks of P.E. beam technology are presented, along with some possible applications for indoor and outdoor security.

Increases in the total market for security devices and the inherent advantages of P.E. beams in certain applications should permit continued growth in P.E. beam sales within the foreseeable future.

INTRODUCTION

Photoelectric beam systems, or "eyes," have been used for more than fifty years, at first for utility functions such as door openers, then to provide security against intruders. They are still in widespread use today, as a result of continued improvements and cost reduction efforts, and their inherent cost-effectiveness and reliability.

Photoelectric beams (P.E. beams) are quite simple and straightforward in operation: light energy transmitted from one end of a protected area is detected at a distance by a photosensitive receiver device. Whenever the beam is broken, the absence of light at the receiver causes an alarm or other action to take place. Lenses and mirrors focus the light at both ends to permit reliable detection at separations up to 300 meters in length.

1983 International Carnahan Conference on Security Technology, Zurich, Switzerland, October 4–6, 1983.

HISTORICAL NOTES

In the very earliest systems, steady, visible light was used. This provided easy alignment, since the installer could see the beam path, especially in a dusty environment. Unfortunately, so could the intruder, who could then avoid the beam or saturate the receiver with a bright light. Sunlight could also easily inhibit correct operation. IR filtering and pulsed light sources provided the first viable systems and met with some success, but the systems still suffered from the relatively short life expectancy of the incandescent bulb, and its high power consumption and heat generation.

It wasn't until the ready availability of the Gallium Arsenide Infrared Light-Emitting Diode (IRLED) around 1970 that an adequate solution was feasible. These devices were designed to radiate only infrared light at about 9000 angstroms, at a much greater efficiency than incandescents. It was now possible to produce a strong, modulated light signal from these small, solid-state devices, which is detectable at great distances, but which is completely invisible to the human eye. IRLED's could also be pulsed to give a very high IR light output for very short durations, making it extremely difficult to cheat with an external light source. This modulation of the light was possible because the IRLED responds to changes in current input at micro-second speeds. These techniques brought some sophistication to the beam technology, and permitted highly reliable detection with low false alarm rate. The power consumption of the transmitters dropped from watts to milliwatts, making these units both more economical and less susceptible to heat problems. In addition, the solid-state IRLED's, when properly used, had life expectancies ten to a hundred times that of incandescents. This meant that reliable operation could be expected for many years without problems with light sources.

The receiver contains optics which focus the light energy to a light-sensitive device such as a photodiode. The received signal is then amplified and used to drive a relay. While the light is being detected, the relay is held energized. If the beam is interrupted for more than a few milliseconds or if power is lost for any reason, the relay is deenergized, causing an alarm. This points out a major advantage of an active system like the photoelectric beam; the system is continuously supervised.

Special plastic materials, through which IR light energy will pass, but which appear completely opaque, also have become available. Devices made of this material prevent the observer from discerning the exact direction of the beam, making the system more difficult to cheat. Anti-tamper switches were also added, which reported an attempt to get at the system, even if the alarm circuit was not activated.

The use of photoelectric beams grew quickly, as did several companies who had worked on perfecting the devices through the cycles described above. However, the P.E. beams still had a few drawbacks which rendered them less suitable for certain applications. Since the P.E. beams rely on narrow rays of infrared energy for detection coverage, they are essentially a "line" or "fence" detector, and require interlaced multiples of transmitters and receivers for high-security space coverage. In addition, the invisible beam made it necessary to use fairly sophisticated techniques for alignment, especially with very long separation distances.

The availability of more advanced alignment tools and special anti-false alarm techniques have kept beam use very much alive. For perimeter coverage or very long separations, photo beams are often by far the most cost-effective. Some beams operate at separations up to 150 to 300 meters. For outdoor use P.E. beams are often set up and checked using a perforated metal plate which obscures 98% of the light beam diameter. This will assure proper operation for badly degraded conditions. In fact, if a high-contrast object can be distinguished at the separation distance, the system should operate without disqualification.

Beam use for space coverage has diminished with the development of motion detectors (PIR, ultrasonic and microwave), especially for indoor use. The passive IR detector, in particular, has captured the lion's share of the increase in security installations. The use of active infrared beams has continued to grow, however, since they have advantages over other technologies in certain applications, such as long distance protection or high turbulence environments.

P.E. BEAM CONFIGURATIONS

Many variations of P.E. beam configurations are possible, from the most basic system of a single transmitter and receiver pair, to those involving multiples of each and mirrors. The following examples illustrate a few of the ways devices are combined to provide security.

Single-ended systems use mirrors or reflectors to bounce transmitted light back to a co-located receiver (Figure 1). This provides short-range coverage at lower costs, with installation of a single unit, but can be cheated inadvertently or deliberately by objects in the beam path, and should be used with caution. A calibrated time measurement technique could be used with a pulsed beam to prevent that cheating. A decrease in the preset time for detection would trigger a trouble signal and/or alarm as required by the application.

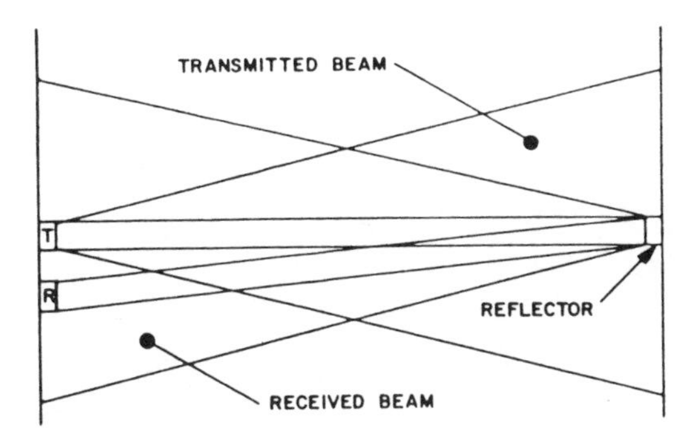

Figure 1. Single-ended system.

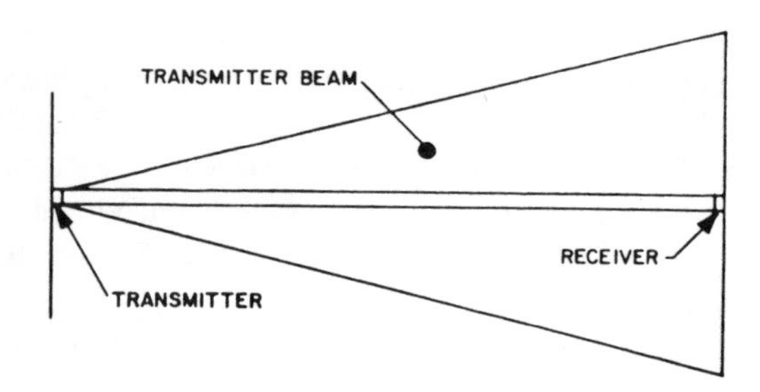

Figure 2. Transmitter beam with single receiver (top view).

The double-ended mode (Figure 2) provides a more secure installation with a single transmitter and receiver pair. If the beam is blocked or power is lost for any reason, an alarm is generated. Stand-by batteries can be used for more reliable power supply.

The use of *multiple receivers* can provide increased "space" coverage or an invisible "fence" (see Figure 3). If any beam is broken, an alarm is generated. In some systems, this is modified to require blockage of multiple beams before an alarm is generated, particularly in outdoor use. This technique permits rejection of interference due to flying debris, birds and small animals, thus reducing the incidence of false alarms.

The addition of *mirrors* (Figure 4) to a system can create a large number of variations and decrease the number of active devices. Caution is required when using mirrors; each one will make alignment more critical and reduce range by 25 to 50%. Movement of a mirror by 1 degree can cause an effective alignment angle shift of 2 degrees. Mirror reflectivity is decreased with increasing angle of incidence, and especially with accumulations of dirt or dust.

Multiple transmitters and receivers can be combined in loops and/or stacks to create highly efficient and dense coverage for perimeters and large spaces (Figure 5).

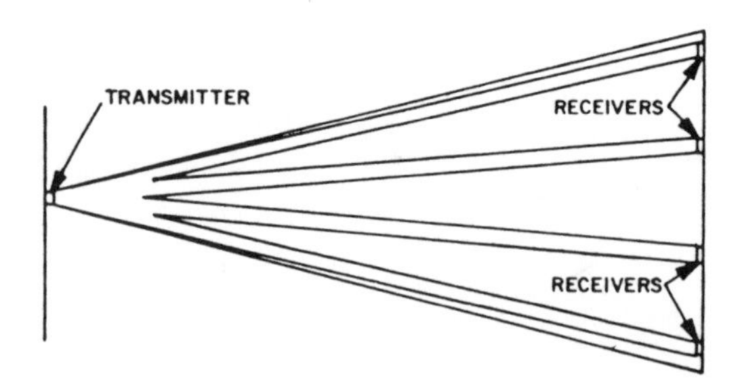

Figure 3. Transmitter beam with multiple receivers.

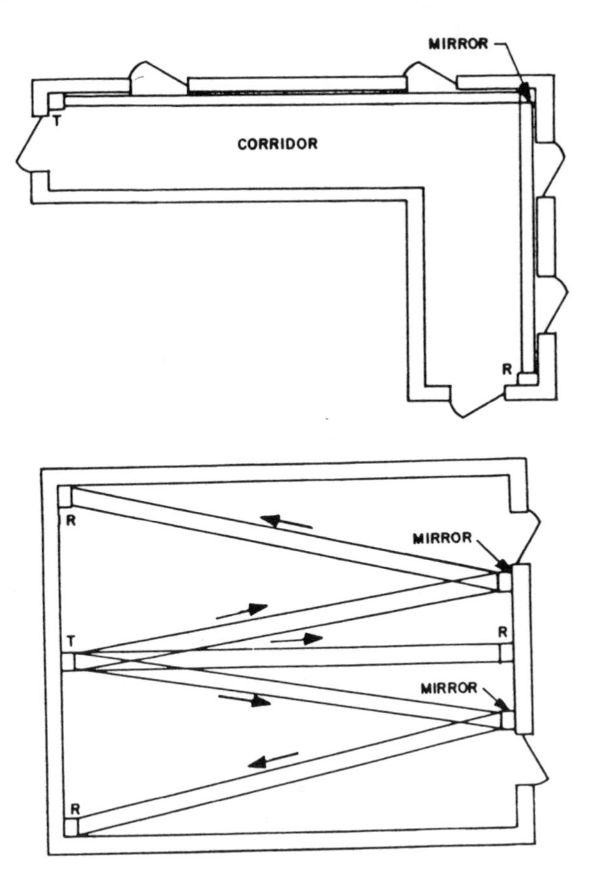

Figure 4. Increasing coverage with mirrors.

A *P.E. beam* can also be used to provide a supervised wireless transmission path (Figure 6). In the simplest case, it can provide monitoring for a detector loop located 150 meters or farther away from the control, without requiring a solid connection. This is also useful for alarm annunciation between remote buildings or other applications where a wired connection would be difficult or expensive.

P.E. beams could also be used for more general communications use. Modulation of the beam could provide reliable and secure supervised wireless transmission of data such as status and identification codes, or even commands, to the centralized control system nearby. There are many possibilities.

For higher outdoor security applications, *multiple P.E. beams* have been used in combination with buried ground-following detector methods, such as seismic or ported coax systems (see Figure 7). The seismic systems use buried dynamic microphones to detect vibrations. Ported coax systems use pulsed or continuous RF energy in one cable whose jacket has holes for controlled radiation. A second similar cable, parallel

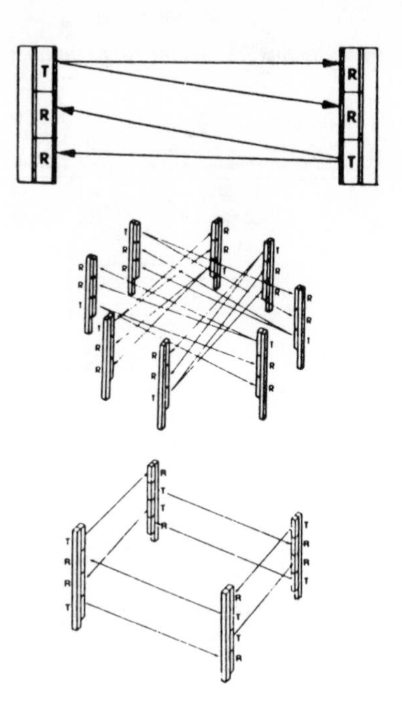

Figure 5. Multiple beam systems.

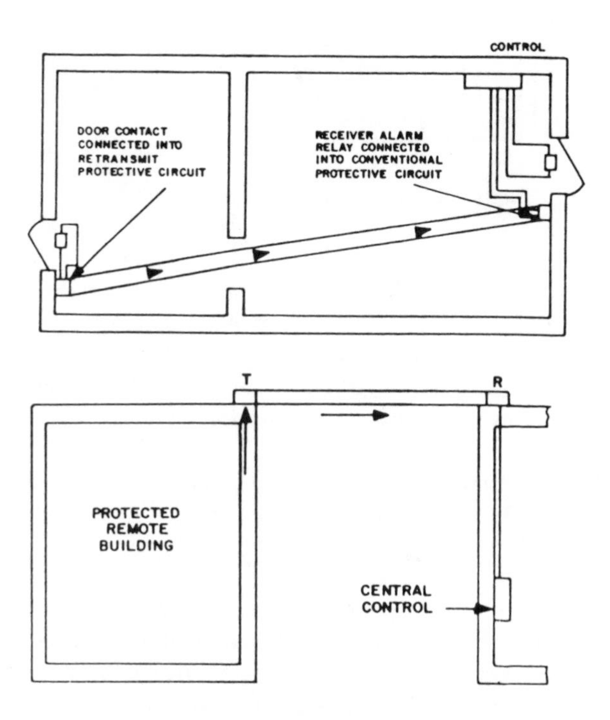

Figure 6. Wireless communication.

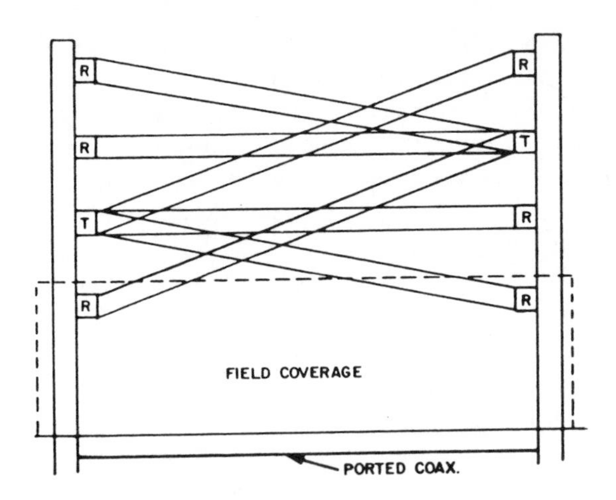

Figure 7. P.E. beams used with ported coax.

to the first, is connected to a receiver which detects changes in the resultant electromagnetic field. These methods would cover the lowest level above the ground, where snow or vegetation might obscure the beam. The multiple beam configuration above it provides a high fence made more difficult to breach without detection.

ADVANTAGES OF P.E. BEAMS

1. Little space (width) is required, since the beams are very narrow.
2. For long distances (greater than 100 meters), they still provide the most cost-effective detection.
3. Excellent perimeter-type protection is provided.
4. Most positive detection technology, highest of present methods.
5. Lowest inherent false alarm rate—relatively insensitive to environment compared to motion detectors. Turbulence, movements outside the beam and loud noises do not affect P.E. beams.
6. Low susceptibility to interference from RF or electromagnetic sources.
7. Alignment is very easy for indoor applications.
8. Systems are fully supervised. If a transmitter fault or power dropout occurs or the transmission path is obscured, the relay drops out to indicate an alarm.

DRAWBACKS OF P.E. BEAMS

1. Straight line or ''fence'' coverage is less suitable for area protection.
2. Care must be taken to keep optics clear and clean in difficult environments.

3. Alignment is often more difficult in outdoor applications due to large distances, ambient light conditions and tools required.
4. In outdoor use, windblown debris, birds and animals can cause false alarms (2-beam blockage or signal processing can help).
5. Fog, rain or snow can cause the beam to be scattered, which could permit an intruder to pass undetected. It can also cause detector disqualification, where that feature is incorporated.
6. Outdoor systems may require heaters to eliminate condensation problems in the optical system.

APPLICATION OF P.E. BEAMS

It would be impossible to attempt an exhaustive list of suitable applications for P.E. beams in security. The sampling which follows should give an idea of the breadth and scope of possibilities for P.E. beam use, which should help to assure its future growth.

Indoor Detection

1. Space protection with multiple beams, receivers and mirrors can be useful in problem areas, where heavy turbulence or other conditions make it difficult for motion sensors. Note: mirrors must be used with a great deal of caution.
2. Reliable counters for conveyer use or any traffic count need.
3. Door and window protection.
4. Corridor protection in factories, schools, etc.
5. Protection of very large areas.
6. Modulated communication paths between detectors, controllers or other devices provides relatively interference-free and secure supervised wireless communications.

Outdoor Detection

1. Perimeter protection for house, business or factory for lower security applications.
2. Combination with ported coax, seismic, microwave, or other technology for high security applications, i.e.,
   a. Power plants and substations
   b. Petroleum refineries
   c. Nuclear power facilities
   d. Nuclear storage facilities
   e. Military installations
   f. Research facilities

   g. Prisons
   h. Large estates and luxury residences
3. Supervised wireless transmission path for alarm signals, especially for remote buildings.
4. Counting vehicles, pedestrians, etc. for any application.
5. Utility uses: gate or door opener
6. Protection of parking lots, rooftops, truck lots, school yards, etc.
7. Provide warning and security/safety lighting for driveways.
8. In areas where windblown debris, birds or animals can cause false alarms, requiring simultaneous breaking of multiple beams or a longer break time will solve most of those problems.

OPPORTUNITIES FOR P.E. BEAMS

Photoelectric beams are very versatile devices, as shown previously in this paper, which will likely find extensive use in the security industry for the foreseeable future. Obviously P.E. beams cannot satisfy every application need, but they are a significant part of the arsenal available to the security installer for complete protection. New technological developments, in electronics especially, provide more stable and higher reliability IR emitters, and detectors as well as other components used in the detectors. Microprocessors and large-scale integration can provide increased features and flexibility in smaller packages at still lower costs. Signal processing techniques will also help to increase tolerance to false alarm sources. Modern plastics permit low-cost, eye-pleasing designs with increased resistance to shock and other environmental conditions. These factors will undoubtedly help to make P.E. beams even more attractive where reliable detectors and low false alarms are a major concern, particularly for outdoor use and perimeter applications.

Total security needs are increasing in the world marketplace, as more people perceive an increase in crime statistics and demand protection. P.E. beams are cost-effective and relatively straightforward in many applications, and should continue to hold their own in the market. In outdoor application, for example, P.E. beam installations are reported to cost about one-third less than microwave per foot of coverage.

The market size in the U.S. alone for P.E. beams in 1983 is estimated at about 150,000. Projections for the next five years show those numbers increasing to over 250,000 units. That is not fantastic growth, but a respectable opportunity, nevertheless, and worthy of attention.

Acknowledgments

Jack Dowling and Bill Kahl, Arrowhead Enterprises, Inc.

REFERENCES

1. H. William Trimmer, *Understanding and Servicing Alarm Systems* (Boston: Butterworth, 1981).
2. ''Product Reliability Survey,'' *SDM*, May 1981, pp. 26–33.
3. ''Finding Profits in Perimeter Detection,'' *SDM*, June 1980, pp. 20–5.

New Developments in Ultrasonic and Infrared Motion Detectors

P. Steiner
P. Wägli

Abstract. In recent years both passive infrared and ultasonic motion detectors became widely accepted intrusion detection devices. In order to arrive at an acceptable false alarm rate, signal processing methods have been developed which made the devices easy to install but also caused a general reduction of their sensitivity.

In this report we present results obtained on the characteristics of typical burglar and false alarm signals. Various methods of ultrasonic and passive infrared signal processing will be discussed with respect to false alarm rejection and detection performance. It was found that algorithms can be developed which give a drastically increased false alarm rejection, thereby maintaining or improving the detection performance of the detector.

INTRODUCTION

Today ultrasonic (US) doppler and passive infrared (PIR)[1] detectors have a wide application as motion detectors in intrusion alarm systems. In recent years signal processing methods have been developed which made PIR and US detectors easy to install and false alarm proof. The major breakthroughs in the PIR technology were the introduction of the pyroelectric sensor, which gives an improved detectivity as compared to the originally used thermistors,[2] and the use of differential infrared sensors[3] which efficiently suppresses false signals caused by air turbulence.

The main step in the ultrasonic technology was the development of the two sideband analysis of doppler[4] signals to reject swinging objects and broadband noise sources (e.g., telephone bells) and thus reject false alarms. However, a general re-

1983 International Carnahan Conference on Security Technology, Zurich, Switzerland, October 4–6, 1983.

duction in sensitivity was tolerated to allow for a wide use of PIR and US motion detectors.

In this report we present results obtained on the characteristics of typical false alarm signals. Methods of US and IR signal processing will be discussed with respect to false alarm rejection and detection performance. It was found that a limitation of the dynamic range of the processed ultrasonic signals can cause a significant loss of important information which leads to a major reduction in sensitivity if doppler signals are evaluated in the presence of environmental noise. It will be discussed how spectral data can be gained and processed to obtain an optimum ratio of sensitivity to false alarm rejection.

Since passive infrared detectors have to detect single events, the alarm has to be triggered at a high threshold to noise ratio such that the sensitivity is limited to a certain minimum temperature contrast and walking speed range of the object to be detected. The use of differential IR sensors can further reduce this sensitivity. It will be demonstrated that the application of signal processing techniques can significantly improve the detection performance of present IR motion detectors.

The implication of these new technologies on future detector designs will be discussed.

PIR MOTION DETECTORS

The schematic of a typical PIR motion detector is shown in Figure 1. The optic focuses the radiation from different fields of view onto the infrared sensor. If a warm target (i.e., human body) crosses this field of view a typical waveform as shown in Figure 2 is produced. The width of the waveform is proportional to the target speed and the amplitude is proportional to both the target speed and the temperature contrast to the background, since the response of the sensor is strongly frequency dependent. This means that a very low amplitude is produced at low and high target speeds due to the thermal and the electrical impedance of the sensor, respectively.

In many applications differential sensors are used as shown schematically in Figure 3. Signals originating from temperature fluctuations of the housing as induced by warm air turbulence are now cancelled since the two differential zones overlap in the neigh-

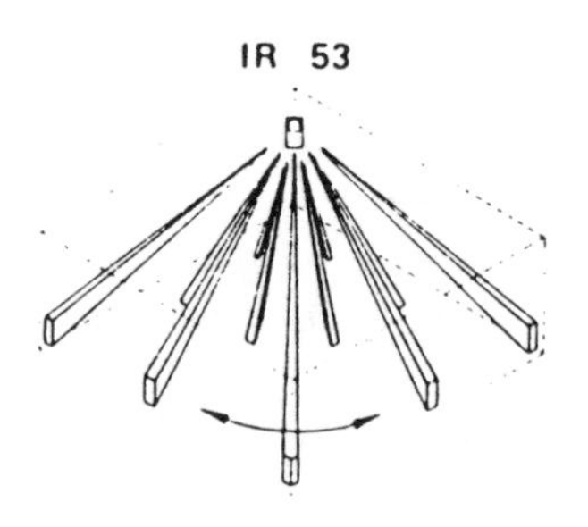

Figure 1. Schematic of a PIR motion detector.

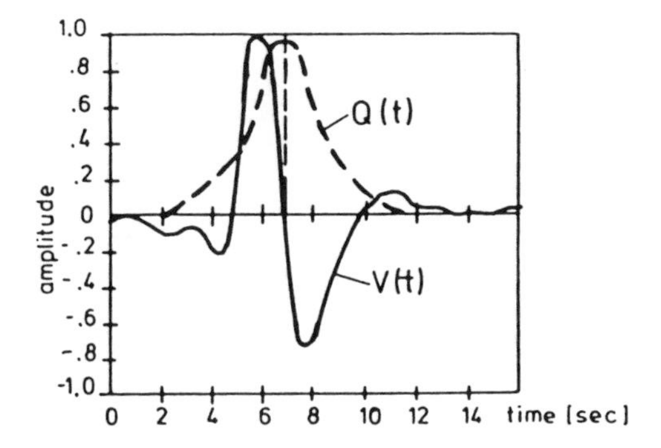

Figure 2. Typical PIR signal waveform Q(t): charge, V(t): output voltage.

borhood of the detector. This also leads to a reduction of the sensitivity close up to the detector. This behaviour is shown in Figure 4, where the signal amplitude measured at the alarm comparator input is plotted versus frequency for a single and a differential sensor.

In conventional systems an alarm is triggered if the amplitude of the signal exceeds a certain preset amplitude. In order to evaluate the performance of a motion detector a computer code was written, which simulates the false alarm behavior and the detection probability of a PIR motion detector. The schematic of the computer code is shown in Figure 5. The detector is characterized by the optic, the sensor and the frequency response of the amplifier. The optic is described by the zone geometry and image quality which allows for the calculation of the infrared radiation intensity as a function of time for a given target size, speed, temperature contrast and distance of the target to the detector. The sensor is characterized by its radiation absorption, pyroelectric coefficient, and thermal and electrical impedance. This, together with the amplifier, allows the calculation of the output signal of the detector electronics.

In order to simulate the detection probability, the computer randomly selects speed, temperature contrast and target detector distance. The output signal is calculated and an alarm is counted if a preset amplitude threshold, a_{th}, is exceeded. This procedure

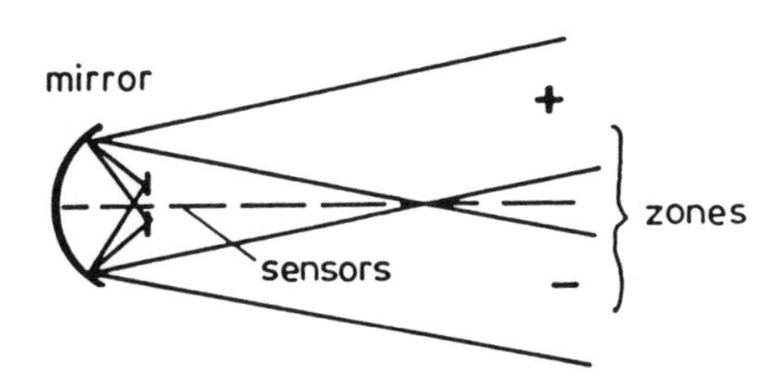

Figure 3. Zone geometry of a differential sensor.

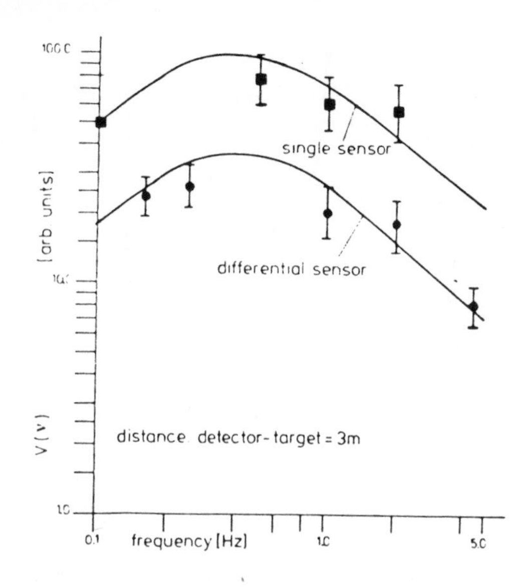

Figure 4. Output voltage of a single and a differential sensor as a function of frequency.

is repeated N times and for a certain value, a_{th}, the number of alarms, $N(a_{th})$, is counted. The detection probability can then be calculated as $p(a_{th}) = \lim_{N \to \infty} N(a_{th})/N$

In a similar way, the false alarm probability is counted. A noise model[5] for the situation to be stimulated (electronic noise, air turbulence, spikes etc.) calculates the frequency power spectrum, a(f), of the noise and a time signal is calculated from that spectrum by a fourier transform with random phases for the different spectral contributions. Again the number of alarms, $N(a_{th})$, is counted for which a preset threshold, a_{th}, was exceeded and the false alarm probability, $p(a_{th})$, is calculated. The result is shown in Figure 6a. The amplitude statistic, p(a), is plotted versus signal amplitude, a. It can be seen that the false alarm rate drops drastically if an alarm threshold/noise ratio of >10 is maintained ($\sim$ 500 mV) in the case of pure electronic noise.

The presence of warm air turbulence as caused by a 2 kW air blower placed 1 m below the detector, however, still gives an excessive false alarm rate. This problem can be overcome by the differential pyroelectric sensor which efficiently suppresses turbulence signals.

The influence of the threshold on the detection is also plotted (Figure 6b) for two cases.

a. A normal burglar was assumed, who does not intend to deceive the system. The following parameter ranges were selected for the attack:
   range : $2 \text{ m} < d < 10 \text{ m}$
   temp. contrast: $2°C < \Delta T < 5°C$
   speed : $0.3 \text{ m/sec} < v < 3 \text{ m/sec}$

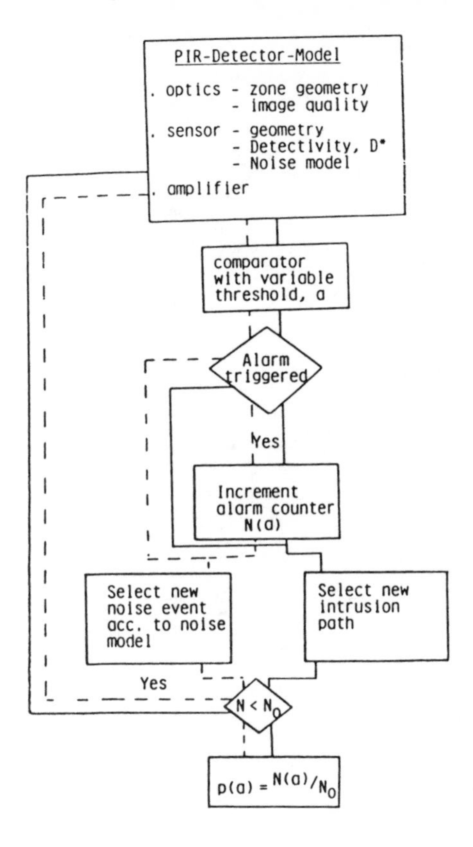

Figure 5. Block diagram of the computer code used to simulate the performance of PIR motion detectors.

b. A burglar with system know-how was assumed in this experiment. A slow speed was assumed together with a low temperature contrast.

range	:	$2 \text{ m} < d < 10 \text{ m}$
temp. contrast:		$0.5°C < \Delta T < 2°C$
speed	:	$0.1 \text{ m/sec} < v < 0.3 \text{ m/sec}$

In all cases $N = 10^3$ attacks were simulated. In case a, PIR motion detectors are good as are reliable intruder alarms; in case b, however, the detection is decreased to a probability of only 30%. This could be improved by lowering the alarm threshold but therefore increasing the false alarm rate. In order to overcome this problem the signal was not only evaluated as a function of amplitude but also the signal shape was measured to determine whether the signal was caused by a target of a certain speed or by random noise. Both criteria are then evaluated and alarm is only triggered if the amplitude exceeds a preset threshold and the signal shape corresponds to a target speed in the preselected range.

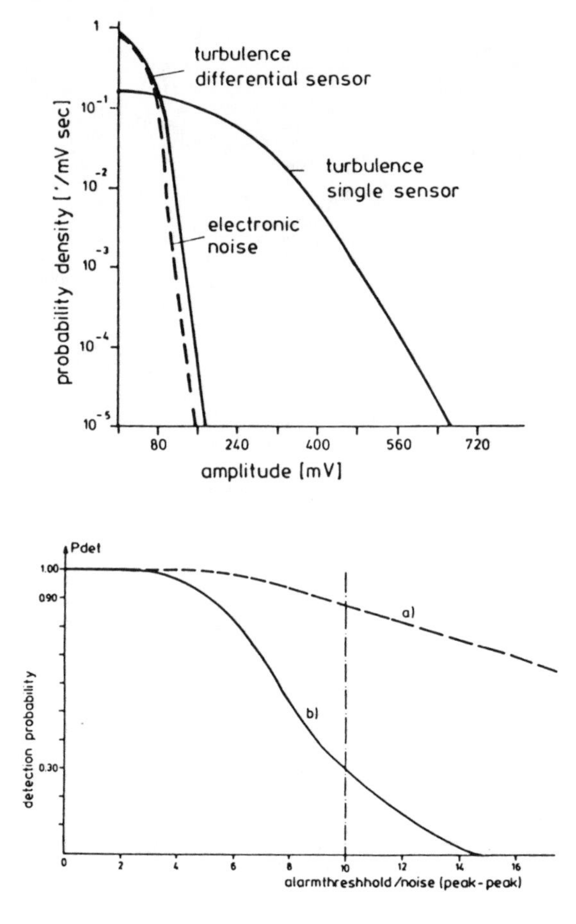

Figure 6. a. Amplitude distribution for signals generated by turbulence and electronic noise for a single and a differential sensor. b. Detection probability as a function of the alarm threshold.

The result is shown in Figure 7. The detection threshold can now be drastically reduced and still maintaining the same false alarm probability, or alternately the false alarm rate can be drastically reduced maintaining the detection probability.

ULTRASONIC MOTION DETECTORS

A block diagram of the most frequently used ultrasonic motion detector circuit is shown in Figure 8. A moving, reflective target produces a spectrum of doppler-shifted frequencies as shown in Figure 9. The maximum occurs at the frequency which corresponds to the target speed, whereas the other spectral components are produced by relative velocities of the moving body parts (i.e., arms, legs, etc.). The amplitude of

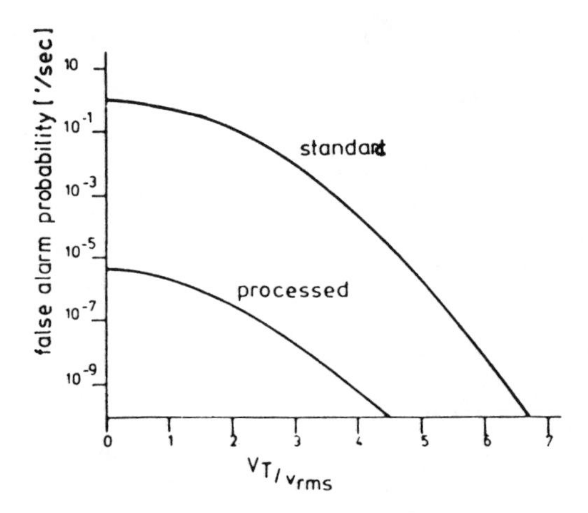

Figure 7. False alarm probability as a function of the alarm threshold for standard PIR detector and for processed signals.

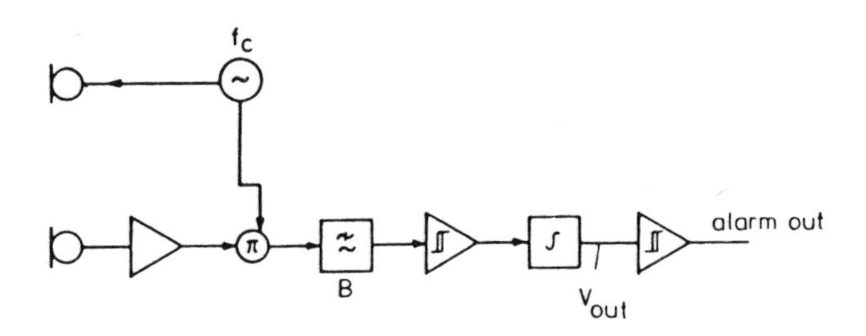

Figure 8. Block diagram of a simple ultrasonic (US) motion detector.

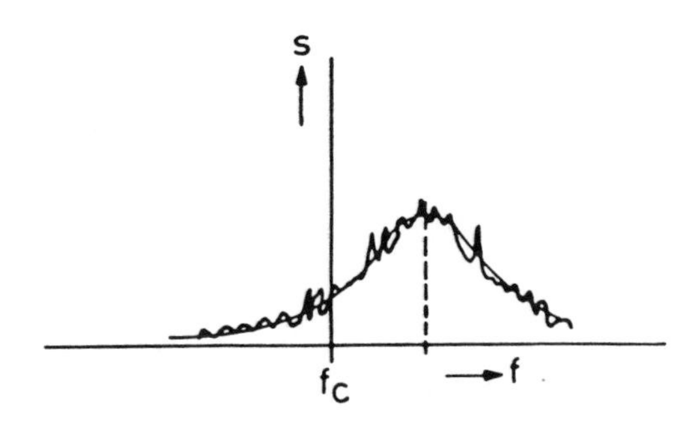

Figure 9. Typical doppler spectrum.

the doppler signal depends only on the distance between the target and the detector, the target reflectivity and the room geometry.

The most simple US detector mixes the incoming signals to a center frequency of O Hz, such that both sidebands occur in the same frequency range. A filter selects the frequency range corresponding to target velocities of 0.1 m/sec $< v <$ 5 m/sec. It is therefore not possible to determine the direction of the movement. The subsequent alarm integrator integrates the filter signal if it exceeds a certain preset limit. For continuously moving targets alarm is triggered after a fixed integration time.

Today this type of signal processing is a standard method, but it is very susceptible to environmental noise. In particular swinging objects (e.g., curtains), telephone bells or air turbulence can create signals which can trigger false alarms.

A common feature of all these disturbance sources is a spectrum with contributions on both sides of the carrier frequency, f_c. Typical signals for telephone bells and turbulence are shown in Figures 10 and 11. The spectrum of a telephone bell consists of almost white noise with distinct resonance peaks at certain frequencies. For the most simple circuit the signal bandwidth is only limited by the bandwidth of detector filter B, and a serious limitation for both detectable range and target speeds occurs.

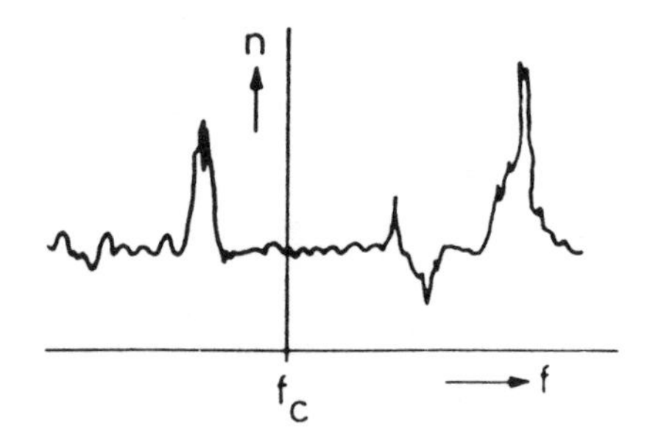

Figure 10. Typical spectrum obtained from a telephone bell.

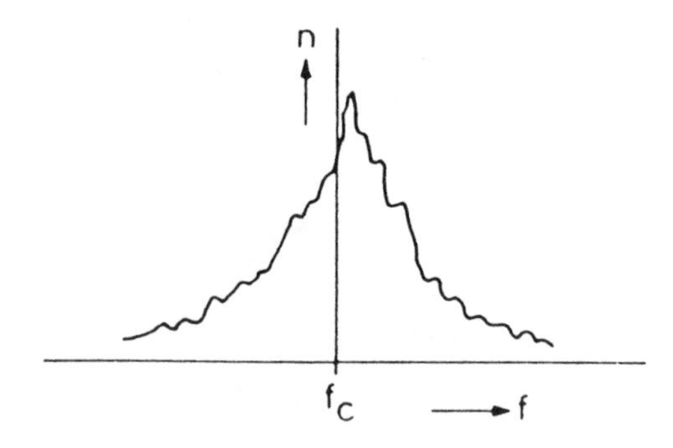

Figure 11. Spectrum obtained from air turbulence.

If noise sources have to be rejected, the bandwidth B has to be narrowed, which results in a narrow detectable speed range and the threshold has to be high, which results in a reduced range. It can, however, be seen that the spectrum is symmetric to a high degree for all the mentioned disturbance sources.

In order to benefit from this situation several US detectors detect not only the amplitude of the doppler signal but also the sign of the doppler shift. If the input to the alarm integrator is multiplied by the sign of the doppler shift the average output of the alarm integrator remains close to zero.

This type of signal processing has, however several drawbacks. Let us first consider a swinging object. Such a target can be easily modeled by the following formula:

$$S(t) = A \cdot \cos\left[t \cdot \frac{2\pi f_c}{c} \cdot \hat{v} \cdot \cos(2\pi f_s t)\right] \quad (1)$$

where f_c Carrier frequency
c Sound velocity in air
f_s swing frequency
v peak swing speed

For typical objects the range for $\hat{v}$ and f_s is 0.3 m/sec $< \hat{v} <$ 3 m/sec and $0.2 < f_s < 5$ Hz, respectively. Since the detector has to trigger alarm for a distance of d ~ 1-3 m for all walking speeds, the integrator time constant is usually in the order of t ~ 1 sec. If now the swing frequency, f_s, decreases for a given speed, v, and gets into the order of the integrator time constant the integration is not over a full swing period and the integrator output can show considerable deviations from zero, which can cause false alarms. This is shown in Figure 11 where the alarm integrator output is plotted versus swing frequency.

This situation can only be improved if the speed of a target is measured and if the integrator integrates with a time constant which is inversely proportional to the target speed, such that alarm integrator output is proportional to the relative distance for which the target moved. This can be accomplished by different methods such as: digital or analog/digital signal processing like DFT, FFT, digital filtering, zero cross counting etc. The improvement obtained by estimating the doppler shift (i.e., the velocity) with only four discrete filters is shown in Figure 12.

In contrast to the first approach the integrator output remains below the alarm threshold even for very low swing frequencies.

In the case of telephone bell noise a similar approach can be used. The signal to noise level with a telephone bell as a disturbance source can be as high as −10 dB for a signal S_d created by a moving person at the upper range limit. In this case an AM-demodulation has to be used since in this region it is clearly superior to a FM-demodulation.[6]

The following approach can be used: filter pairs are arranged symmetrically around the carrier frequency, and the difference voltage, v_d, of the output of the filters is calculated. For a noise input the output fluctuates with time whereas the fluctuation

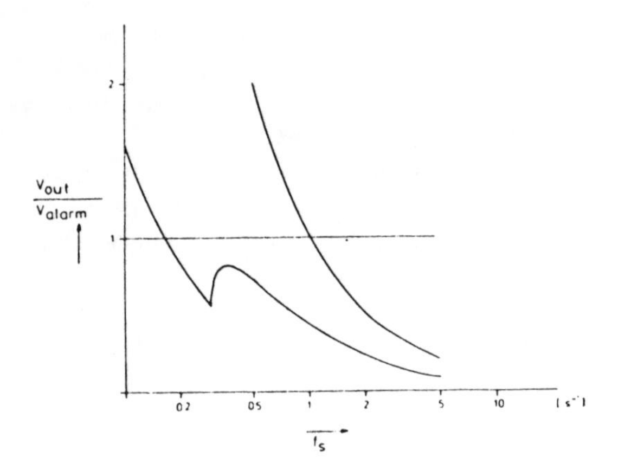

Figure 12. Output of the alarm integrator in the presence of swinging objects a. without a measurement of speed, b. with a measurement of speed.

amplitude is proportional to the variances of both filter outputs. The variance, however, decreases as the observation time increases, which means that even very high noise levels can be rejected if the observation time is long enough.

The noise rejection as a function of observation time is plotted in Figure 13. Also shown is the limiting observation time, t, which is given by $t = d/v_{max}$, where d and

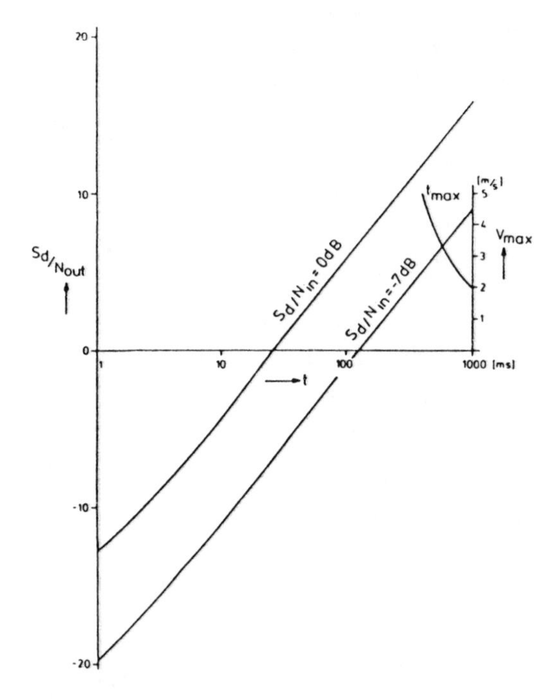

Figure 13. Filter output noise as a function of observation time. S_{d}. = detector threshold level, t_{max} = relative alarm distance/speed, v_{max}.

v_{max} denote the detection distance and the target speed, respectively. It can be seen that a drastic reduction of broadband noise can be achieved and a target can be detected even in the presence of high environmental noise.

This demonstrates that even simple models to simulate the detector performance lead to solutions which clearly improve the US motion detector performance. Speed estimation gives a constant relative detection distance and efficiently suppresses swinging objects, whereas the evaluation of the amplitude difference of filters symmetrically arranged with respect to the carrier frequency significantly reduces the influence of broadband noise.

CONCLUSIONS

In conclusion it has been demonstrated that signal processing technologies can significantly improve the present performance of motion detectors. Since intrusion alarm systems are becoming more and more popular, a new generation of devices with a significantly lower false alarm rate is needed.

In this paper we have shown how new technologies can be used to achieve this challenging goal.

REFERENCES

1. P. Wägli and J. Muggli, *Proceedings of the Journees d'Electronique,* Lausanne, 1982, p. 231.
2. E.H. Putley, *Topics in Applied Physics,* vol. 19, 71 (New York: Springer-Verlag, 1980).
3. F. Schwartz, *Aerospace Technology,* 11 March 1968, p. 30.
4. A.A. Galvin, US-Patent No. 3 665 443, 1972.
5. J. Muggli and P. Wägli, *Proceedings of the SPIE Conference on Advanced Infrared Technology,* Geneva 1983.
6. M. Schwartz, *Information Transmission Modulation and Noise* (New York: McGraw-Hill, 1980).

Vault Protection with Seismic Detector Systems

Kurt Gugolz
H. Eugster

Abstract. Seismic detector systems take advantage of the fact that all tools used by burglars in their attempt to gain access into a vault create sound waves propagating through the wall usually constructed of steel, reinforced concrete or equivalent material. Solid-borne sound waves may be detected by piezoelectric sensors up to a given distance from the point of generation. The range of detection depends on factors such as frequency and amplitude of the generated sound, the attenuation by wall material, the response characteristic of the piezoelectric sensor and the sensitivity of the associated amplifier network.

Apparatus and test methods have been devised to evaluate the quality of walls and measure their attenuation factor. Its configuration and application are described together with other methods of periodic supervision of the correct functioning of individual detectors or comprehensive systems.

INTRODUCTION

Modern sophisticated methods of attack on vaults call for an efficient security system. Qualities most frequently asked for are high reliability, and high sensitivity together with high immunity from false alarms. As criminals utilize new tools to reach their goal, methods of detection must be adapted to this new situation. Elaborate systems must also offer a self-check for local or remote operation.

FORMS OF ATTACKS TO VAULTS

A protection concept for vaults meets requirements if a criminal is apprehended before he succeeds in gaining access to the valuables and before having caused substantial

1983 International Carnahan Conference on Security Technology, Zurich, Switzerland, October 4–6, 1983.

damage. Vault constructions offering physical resistance to break-in attempts for a considerable lapse of time are commonly made of reinforced concrete or equivalent homogeneous material.

Tools used by criminals with the purpose of gaining access to vaults can usually be found on most construction sites: explosives, cutting discs or diamond head drills, oxyacetylene cutters and thermic lances. Each of these tools creates sound waves which use the wall structure as medium. Their rapid propagation offers an ideal means of information transmission to some distant collection point in the wall. An adequate method must be found to transform the mechanical information into electrical signals. Signal discrimination and evaluation could then be performed in electronic circuits within the required period of time. However, solid-borne sound waves emanating from tools cited above differ greatly in amplitude, frequency composition and duration. Higher frequencies, in addition, suffer from higher attenuation along their path to the transducer. Two representative diagrams are shown in Figure 1 and Figure 2 below.

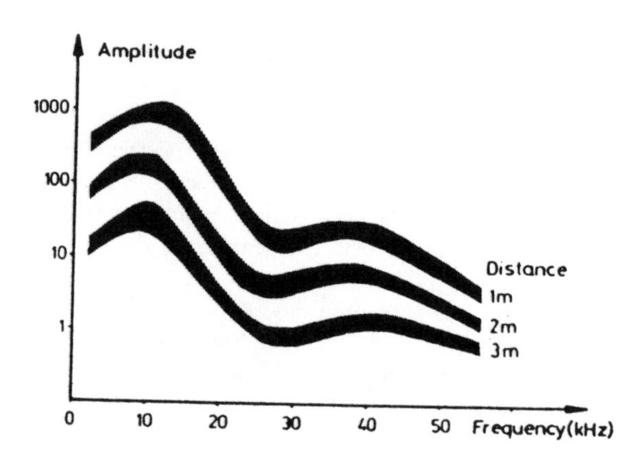

Figure 1. Attack with drill or parting off grinder.

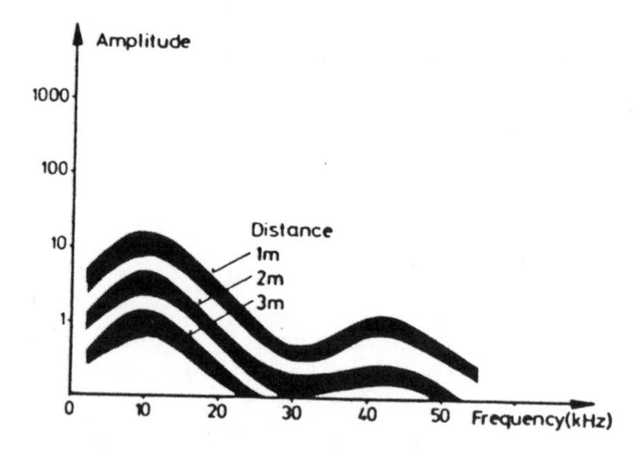

Figure 2. Attack with thermic lance or cutting torch.

The frequency range suitable for evaluation in an electronic alarm system is situated between 7 kHz and 30 kHz. Lower frequencies may stem, for example, from environmental disturbances caused by public transport in the neighborhood of the vault. Higher frequency signals are not of interest as they suffer from excessive attenuation or are in any cause unwanted as they represent airborne noise.

Signals generated by the tools cited earlier are either of high amplitude and short duration (explosion) or small amplitude and long duration (thermic lance) or lie somewhere in between.

SEISMIC DETECTOR CHARACTERISTICS

An electronic system designed for the processing of such a variety of signals is the seismic detector. Its transducer is a piezoceramic microphone having the property of transforming a signal e.g., of 1 mg acceleration into an electrical signal of 0,2 mV. Amplifier, filters, an integration circuit and an alarm output stage form a cost-effective, self contained detector. The seismic detector, due to its outstanding discrimination of environmental disturbances and its high sensitivity, has gained a good reputation for protecting not only vaults, but also safes and Automatic Teller Machines. If certain conditions prevail, it may even be applied to detect attacks on safes used in night depositories. A functional diagram of a typical seismic detector is shown in Figure 3. The detector's response characteristics with the relationships among amplitude, frequency and time up to alarm triggering are shown in Figure 4.

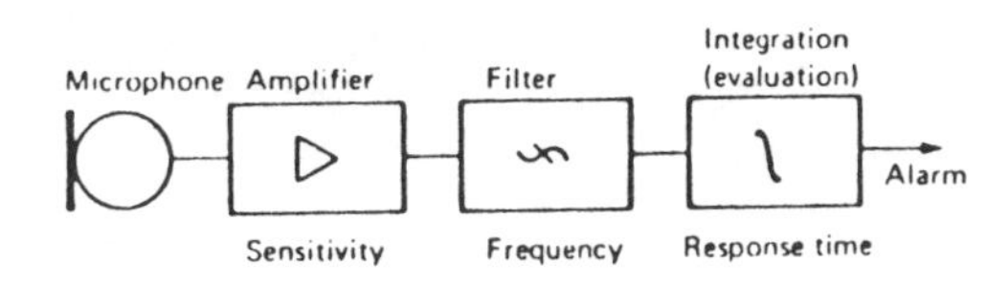

Figure 3. Functional diagram of seismic detector.

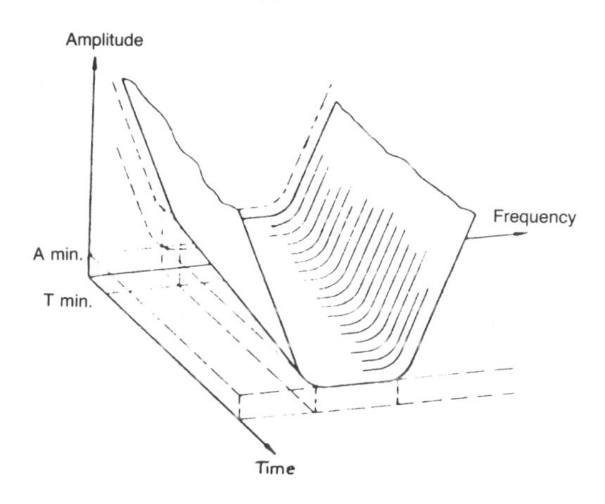

Figure 4. Relationship between amplitude, frequency and time.

Such an arrangement gives the standard seismic detector a range of approximately 3 m radius from its location on the wall. This range is determined by the ability of the device to detect low amplitudes generated by a thermic tool. Although a well-designed detector of this kind gives a high degree of immunity from environmental disturbances, care must be taken with regards to interference sources which can generate signals not suppressed by the filters, i.e., frequencies in the band of 7 – 30 kHz. In order to isolate such interference sources, sensitivity can be reduced and the number of detectors increased.

Characteristics of Seismic Detectors for Special Applications

More sophisticated detectors for special applications use mixers for improved filter characteristics and an automatic integrator reset circuit. The prefiltered sensor signals are mixed with the output of a sweep oscillator in a ''Norton'' operational amplifier. Only the difference frequency is amplified further and passed through a narrow band pass filter of approximately 100 Hz. Thus, signals above and below the highest and lowest oscillator frequency are cut off effectively by the excellent rise and fall characteristics of such a filter arrangement. It allows lowering the upper edge of the filter to a frequency where air-borne sound, for example, from ultrasonic surveillance, is no longer present. Nevertheless the range of the device may be extended by increasing its sensitivity. Typically, the range increases to approximately 4m radius. Furthermore, continuous or intermittent disturbances with narrow bandwidths are eliminated by the use of a relatively low sweep frequency and an automatic integrator reset. This circuit resets the integrator to its initial condition as soon as a few seconds have elapsed without any incoming signals.

This type of detector may be applied cost effectively if disturbances such as noise from spokes of car tires in car parks above vaults, noise from water pipes or ultrasonic sound has to be circumnavigated while still maintaining high sensitivity. In principle, such a circuit is shown in Figure 5.

Seismic detectors may therefore be applied to steel safes without any restrictions. Similarly to corners in vaults, edges on steel safes attenuate sound waves. This fact cannot be neglected when planning the location of the detectors.

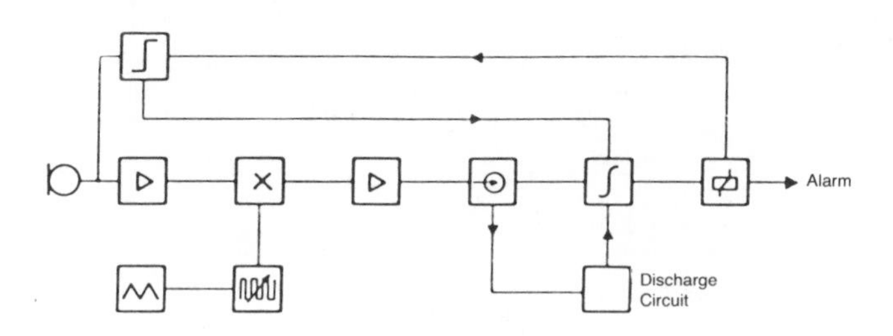

Figure 5. Automatic integrator reset circuit.

APPLICATION OF SEISMIC DETECTORS TO AUTOMATED TELLER MACHINES

If, however, seismic detectors have to be applied to the safes of Automatic Teller Machines (ATMs) care has to be concentrated on the inherent disturbances generated each time the *money-dispensing* mechanism is operated. Unfortunately, there is no typical noise spectrum as each manufacturer of ATMs uses a different type of mechanism to actuate transport and shutters. Typical sound diagrams of two different machines computed using the most advanced detection and recording techniques are shown below in Figure 6.

Seismic detectors which can cope with the task of protecting the safes on ATMs of various manufacturers were developed after detailed research work using advanced measuring techniques, which are also being applied in the development of state of the art test gear for both detectors and walls.

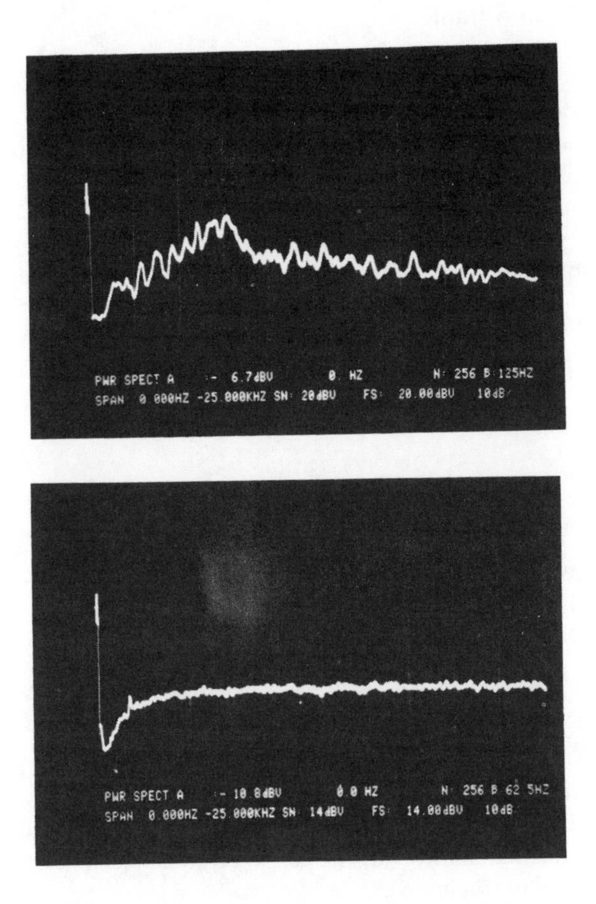

Figure 6. Typical sound diagram of two different Automatic Teller Machines using the most advanced detection and recording techniques.

TESTS AND TRIALS

Testing the Seismic Detector

Transmitting weak vibrations from the wall to the base plate of the detector, and further to its transducer, with a minimum of loss is vital. Hence an on-site check of these joints together with the functioning of the associated electronic circuitry is desirable. This can be realized using a piezoelectric transducer fitted in an appropriate housing and fixed to the wall near the detector. A generator feeds the test transducer with an adequate electrical signal. Frequency and amplitude of this signal are arranged so that, when transformed in the test transducer, realistic vibrations as created by a burglar tool are simulated. These vibrations are sensed and evaluated by the seismic detector which transmits the result (test alarm) to an indication panel.

Testing the Walls of a Vault

Solid-borne sound waves are attenuated as they propagate through a wall. The attenuation by a homogeneous concrete wall is known, hence the range of a detector with a known sensitivity can be determined. However, fissures or heterogeneous material modify the attenuation so that the range can no longer be predetermined. A device using a non-destructive measuring method has been conceived which allows the computation of a set of graphs related to the quality of a wall and the known sensitivity of a given detector. With this information, optimum positioning of detectors is possible (Figure 7).

The system uses a transmitter head, T, which applies vibrations to the wall. Coupling between head and wall is realized with a nondestructive medium. The frequency/amplitude mix simulates the weak signal from a thermic lance.

$$a = \frac{20 \log \frac{p_2}{p_1}}{1}$$

where a = attenuation (dB/m)

p_1 = level of acceleration at receiver R_1 (mg) *

p_2 = level of acceleration at receiver R_2 (mg) *

1 = test distance (m)

$* = 1g = 9{,}81\frac{m}{s^2}$ (gravity)

The two microphones R_1 and R_2 measure the acceleration levels related to the test distance, 1. The electronic circuits of the device process the values such that the characteristic attenuation figure (dB/m) for a given object (wall) may be directly indicated at the display (Figure 8).

From the set of graphs computed, optimum location of detectors with a known sensitivity may be determined (Figure 9).

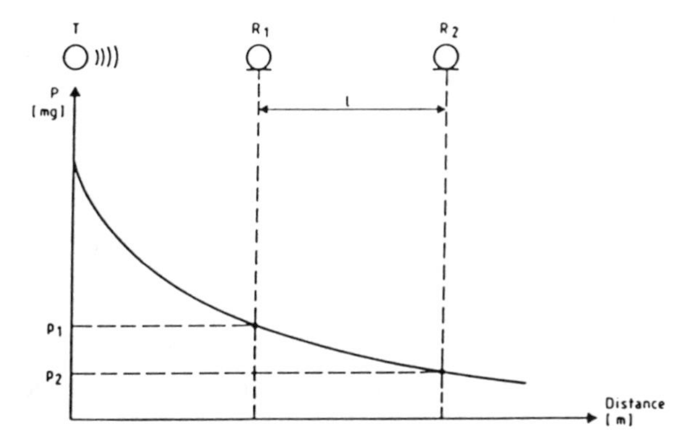

Figure 7. Principle of measurement to determine optimum positioning of detectors.

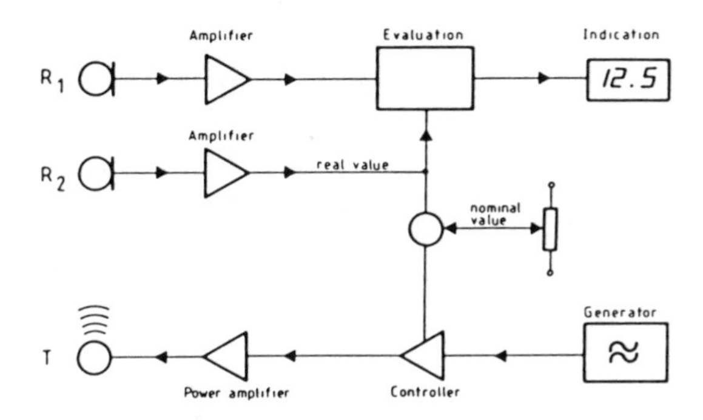

Figure 8. Block diagram.

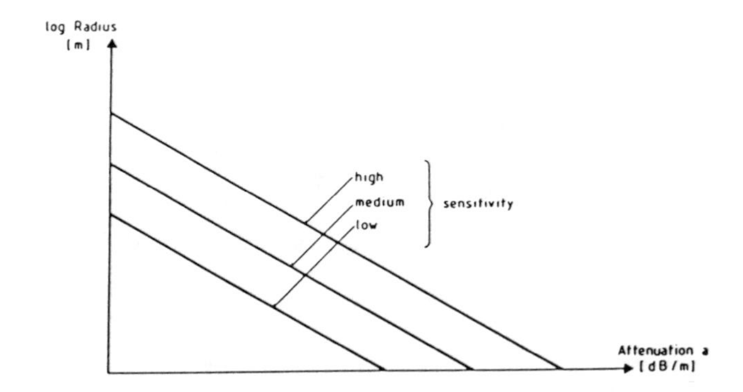

Figure 9. Computed optimum location of detectors with a known sensitivity.

Since the application of both the transmitter and receiver heads are non-destructive, the device has a high degree of mobility.

The Use of Trials under Actual Operating Conditions

The structure of a vault, taking into account corners, doors and other heterogeneities, is very complex. Detectors and test methods developed on a theoretical basis must therefore undergo extensive trials with the aim of proving their reliability in real conditions. In order to be realistic, these trials must not only make use of tools currently favored by criminals, but also be carried out under actual environmental conditions, e.g., in the immediate vicinity of railway tracks.

This type of investigation also provides basic information used for application and installation notes for the benefit of installers and operators of security systems.

Other methods of vault- and safe-protection comprise capacitive techniques, whereby an electric field created around an object allows the detection of a human body when it approaches closer than approximately 30 cm. Surface protection is still in use and so are microphone systems reacting to audible sound within the surveillance zone. The list can be extended with space surveillance, ultrasonic- or microwave doppler systems as well as passive infrared detectors. None of these methods meet the requirements which a seismic detector can fulfil. Some of these are listed below and compared in graphical form with a choice of other detection methods (Figure 10).

The number of installations, a great part in the banking sector, using seismic detectors is increasing every year and thus proving their superiority over other methods.

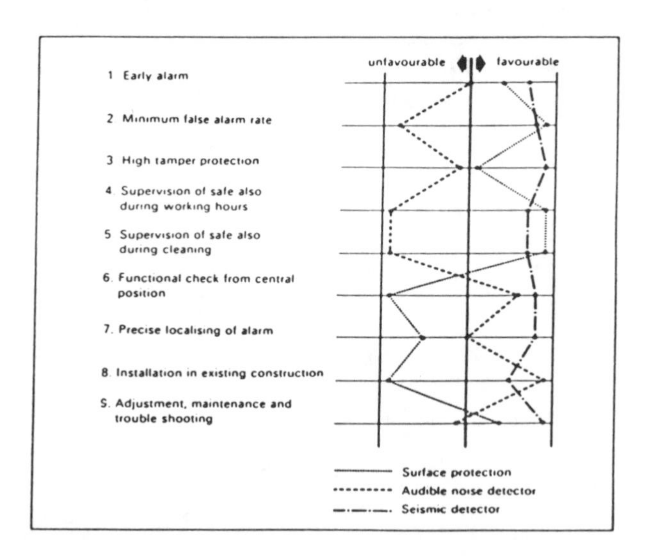

Figure 10. Comparison of seismic detectors and other detection methods.

REFERENCES

Günther, Hansen, Veit, *Technische Akustik* (Ausgewählte Kapitel: Expert Verlag).

J. Tichy/G. Gautschi, *Piezoelektrische Messtechnik* (Berlin, Heidelberg, New York: Springer Verlag, 1980).

Brüel and Kjaer, *Piezoelektrische Beschleunigungsaufnehmer und Ladungsverstärker, Theorie und Anwendung* (Firmenschrift, 1979).

Cremer Hechl, *Körperschall* (Berlin, Heidelberg, New York: Springer Verlag, 1982).

PART IV

External Use of Closed-Circuit Television

Among other uses, external closed-circuit television (CCTV) is employed to interface with perimeter barriers and intrusion detection systems. CCTV, however, can also be used in lieu of barrier and intrusion detection systems.

CCTV systems are used to replace or minimize the use of human patrols. Human patrols are expensive. It is estimated that it costs $150,000 to staff one post 24 hours a day, 7 days a week. The cost for any post increases from year to year with pay raises and inflation. The cost of most technology is much less and the cost decreases with time. Once a system is purchased, the maintenance and replacement become the only cost.

CCTV systems may not only be cheaper, they may be more effective as well. Low-light cameras strategically located can provide much better images at night than a person (even when equipped with night vision binoculars) can receive directly.

The first article, by Kravets, shows the enhancement of CCTV, mandated by the National Regulatory Commission, as a component in perimeter defense. More importantly, it shows the increased benefit of an innovative configuration of cameras over the present configurated methods. By a sophisticated but simple layout, the quality of surveillance is improved at considerable financial savings. This system, as presented, can be applied to most closed and open perimeter facilities to provide financial savings in exterior CCTV systems.

The second article reports on the research results of using low-light TV cameras in lieu of other intrusion detection devices and patrols to detect illegal aliens entering the U.S. across the Mexican border. In addition to addressing the improvement of detection, the project deals with the configuration problems and person-security product interface problems. This system has an excellent potential in a variety of settings.

Artificial light is necessary since intruders typically attack under the cover of darkness. For direct and CCTV surveillance, artificial illumination requires energy and the cost of the energy can be great for large compounds.

In addition, the facility's unique configuration of lamps can aid an enemy by providing easy identification from aloft. If the lighting could be considerably decreased it would make such identification difficult, if not impossible. The third article in this section demonstrates that the strategic use of reflectors greatly lessens the need for energy-consuming lamps while still providing sufficient illumination for CCTV surveillance.

Many security devices increase the complexity and costs of security systems. The article in this section discusses a simple device which greatly improves surveillance ability while at the same time it greatly reduces the cost and complexity of artificial illumination.

Various Concepts on Closed-Circuit Television (CCTV) and Camera Layout for a Power Plant Security System

Zol Kravets

Abstract. This article analyzes various concepts in closed-circuit television (CCTV) camera layout for protected areas and isolation zones of a power plant security system. Monitoring and video switching methods associated with camera layout are described, and an overview of a cost impact for different CCTV systems is provided.

Emphasis is placed on the advantages and cost effectiveness of using an opposed-view camera concept with a double-switching monitoring method, in lieu of one-way orientation with single-switching monitoring for CCTV surveillance of multi-corner perimeters.

Although this paper addresses the problems and solutions in CCTV design for a power plant security system, the concept and analysis discussed also can be utilized in the design of a CCTV system for an industrial, commercial, or military installation requiring perimeter protection with CCTV assessment.

INTRODUCTION

A closed-circuit television (CCTV) system is one of the major components of a nuclear power plant security system, providing a security operator the capability to assess by video any activity taking place on external protected areas or internal vital areas. This paper addresses the application of CCTV for perimeter and protected areas' assessment only, the cost of which is estimated to be from 10 to 15 percent of the total cost of a power plant security system. This includes equipment and installation cost and from 10 to 20 percent of the total maintenance cost. Therefore, the considerations for CCTV

1984 Carnahan Conference on Security Technology, University of Kentucky, Lexington, Kentucky, May 16–18, 1984.

have a significant effect on the overall system cost incurred during construction and maintenance.

BASIS OF DESIGN

CCTV system design concepts for a nuclear power plant are based on Nuclear Regulatory Commission (NRC) requirements and recommendations.

The NRC determines and specifies the areas of the site to be observed and their size:

> Isolation zone shall be maintained in outdoor areas adjacent to the physical barrier at the perimeter of the protected area and shall be of sufficient size to permit observation of the activities of people on either side of that (physical) barrier in the event of its penetration (373.55.C3).[1]
>
> An acceptable security program would typically extend isolation zones at least 20 ft on either side of the protected area barrier (Section 4.1).[2]

In reference to the system capability, the NRC recommends that the fixed CCTV system includes a field of view for the entire isolation zones (both sides of the PA barrier) (Section 6.2).[2]

For the camera layout in the perimeter area, the NRC recommends that where possible, cameras need to be placed so that they are pointing parallel to the adjacent perimeter of the protected area (Section 9.1.1).[3]

In addition to basic NRC requirements and recommendations, the following factors are taken into consideration in the process of designing a CCTV perimeter layout:

1. Any blank spots in the isolation zone should be excluded, and the capability for 100 percent coverage of the isolation zone provided.
2. Human engineering factors involved with monitoring of CCTV cameras should be in an order which eliminates any confusion for the security operator and provides him easy recognition of the area under assessment.
3. Cost-effectiveness of the equipment, installation, and maintenance is directly related to the number of cameras being used.

CAMERA LAYOUT

The layout of CCTV cameras depends on the scene to be viewed and the field of view to be covered. For this application, different lens sizes are employed to satisfy field-of-view requirements with the use of a minimal number of cameras. Figure 1 illustrates field-of-view coverage for a 40-ft-wide isolation zone employing standard fixed lenses for cameras with 1-in. format camera tube and 0.75 percent resolution threshold. This resolution is sufficient for a security operator to recognize a human presence or to detect a small animal. The horizontal angle of view decreases from 53.9 degrees for a 12.5-mm camera to 7.3 degrees for a 100-mm camera.

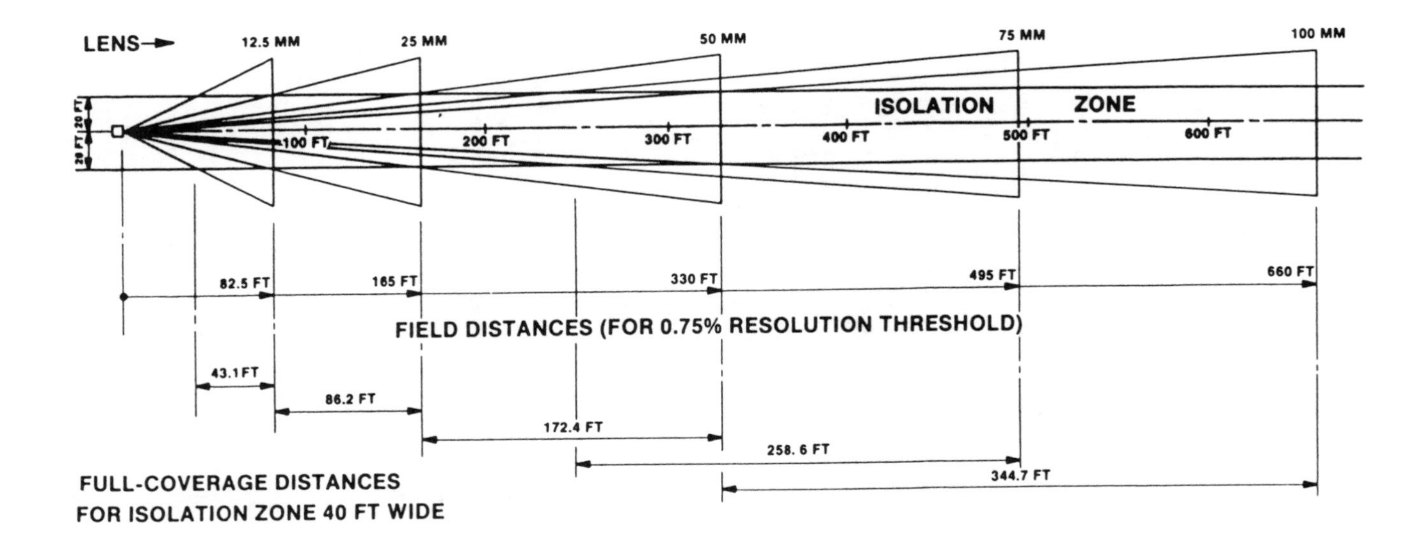

Figure 1. Horizontal field-of-view dimensions.

The blank-spot areas adjacent to camera locations relatively increase from 39.4 ft for a 12.5-mm camera to 315.3 ft for 100-mm cameras. This is a major factor that should be taken under consideration in camera layout.

Presently, two major methods are used in CCTV camera layout:

1. Cameras oriented along the fence in one direction (usually clockwise).
2. Cameras aimed at the fence from the protected area.

Primarily Method 1, or a combination of Methods 1 and 2, is used with cameras aimed at the fence supplementing cameras oriented along the fence to cover any blank-spot areas. This technique applies specifically to nuclear power plant perimeters with an average of from 5,000 to 7,000-ft lengths and multicorner breaks in the length. Such a multicorner perimeter does not allow use of sufficiently long lenses (75-150 mm), due to the resulting long blank spots which have to be covered with supplementary cameras. An example of this is shown in Figure 2.

Two 25-mm cameras should be added to the corner area to cover the blank area caused by the use of two 50-mm cameras for the adjacent two sides of the perimeter.

The use of cameras with short lenses (12.5 mm to 35 mm) provides coverage in the corner area without the use of supplementary cameras aimed at the fence, but dramatically increases the number of cameras along each side of the perimeter.

Figures 3 and 4 show two different camera layouts for a power plant perimeter with a total length of 5,450 ft and 9 corners. In Figure 3, only the camera with a 25-mm lens is utilized; therefore, no supplementary cameras are required to cover blank

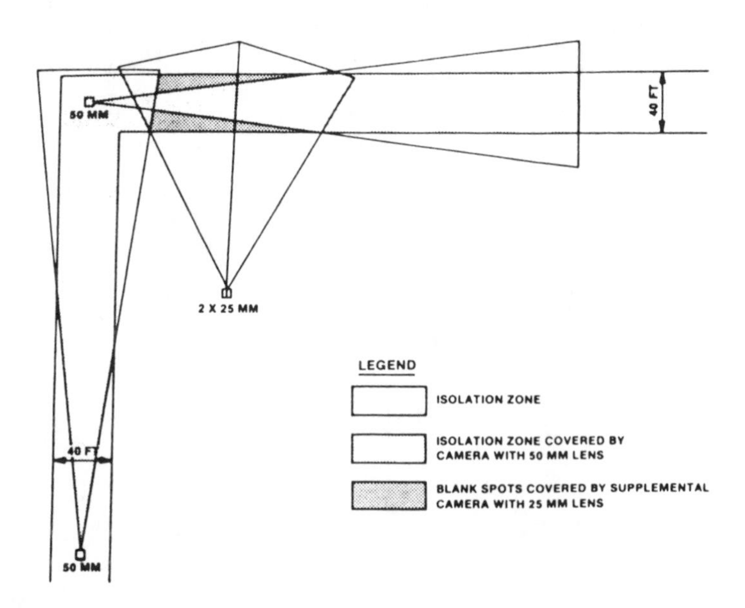

Figure 2. CCTV layout in corner area (detailed).

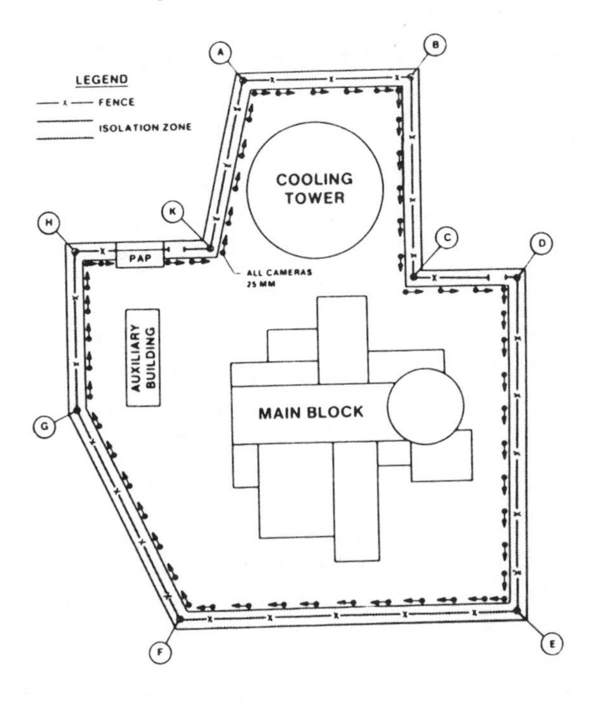

Figure 3. CCTV layout—the camera with a 25-mm lens is used—(Method A).

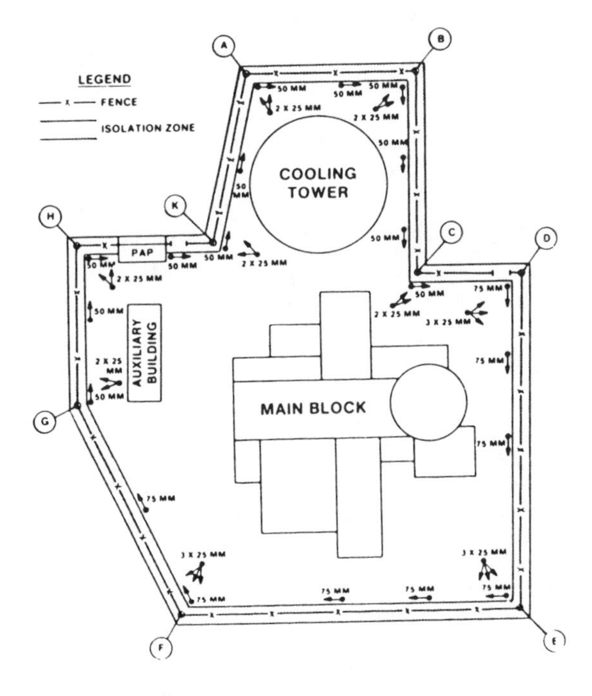

Figure 4. CCTV layout—cameras with 50- and 75-mm lenses are used—(Method B).

spots in the corner areas. In Figure 4, cameras with 50- and 75-mm lenses are utilized with a supplementary 25-mm lens camera to cover corner areas. The results of the computation for the foregoing layouts are provided in Table 1, which addresses the following methods:

- A—One-way orientation with 25-mm cameras all around.
- B—One-way orientation with 50- and 75-mm cameras for the perimeter and 25-mm cameras to cover corner areas.

Although Method B is more cost effective than Method A (fewer cameras and towers for the same application), it is not more effective. In addition, the sequential camera switching and the operation of this system is more complicated due to the disorientation of the security operator when different numbers of supplementary cameras in corner areas switch from the short field of view directed at the fence to the long field of view oriented along the fence.

To resolve this problem, an opposed-camera layout has been evaluated and recommended. This concept, when used properly in combination with appropriate video switching, can provide a cost-effective system that satisfies operational (human factor) requirements. The opposed-camera layout is presently in use in limited application for local areas where none of the other methods can be used. However, for the purposes of this paper, the application of this method is presented as a concept for the entire system, and the video switching has been changed to suit this concept.

To utilize the opposed-camera concept, CCTV cameras with long lenses (50mm to 75mm) are placed in the corners of the perimeter facing in opposite directions, so that two opposed cameras can cover the area equal to the full-lens, field-of-view distance. Each camera covers the blank spots of the opposed camera adjacent to the camera location. Therefore, as two 75-mm cameras can cover an isolation zone with a total length of 495 ft without blank spots, in this case two 50mm cameras can do the same job for a 330-ft length.

This approach permits utilization of a full-camera capability for perimeters with a limited side length (from 200 ft to 1,000 ft) and multiple corners by placing cameras in the corners. Therefore, two cameras can be mounted on one tower and oriented along the adjacent sides of the perimeter.

Table 1 Computations for Alternative Layouts

		Side	*AB*	*BC*	*CD*	*DE*	*EF*	*FG*	*GH*	*HK*	*AK*	*Total*
Method	*Cameras*	*Length (ft)*	*500*	*600*	*300*	*1,000*	*1,000*	*700*	*450*	*400*	*500*	*5,450*
A	Cameras with 25-mm lens		5	6	3	11	11	7	5	4	5	57
(Fig. 3)	Towers		5	6	3	11	11	7	5	4	5	57
B	Cameras with 75-mm lens		—	—	—	3	3	2	—	—	—	8
(Fig. 4)	Cameras with 50-mm lens		2	3	1	—	—	—	2	2	2	12
	Cameras with 25-mm lens (supplementary)		2	2	2	3	3	3	2	2	2	21
	Towers		3	4	2	4	4	3	3	3	3	29

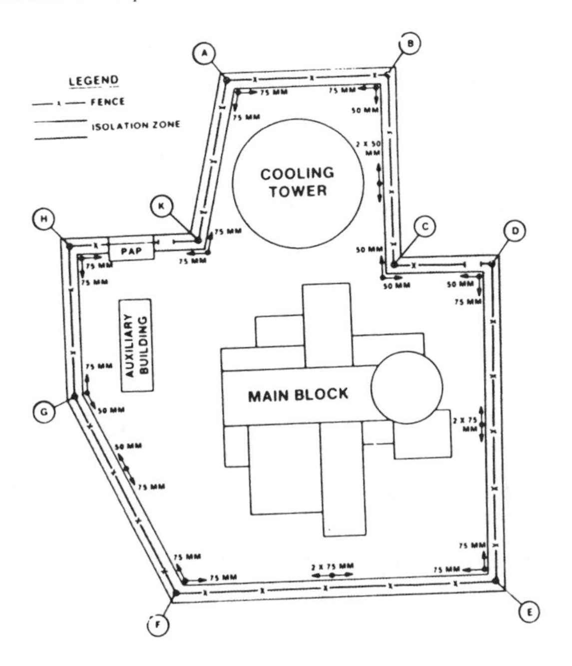

Figure 5. CCTV—the opposed-camera concept—layout (Method C).

The camera layout which employs this method is shown in Figure 5, and the computation of the number of cameras required for each side of the perimeter is provided in Table 2 (see Method C).

For this camera layout, a double video switching method can be utilized. This method provides sequential switching of two opposed cameras simultaneously on two adjacent monitors on the operation console, or so-called zoning switching when one entire zone of the perimeter appears on the two monitors in an opposed direction, then switches to the next zone, and so forth. For holdup, two main monitors on the operator console can be used so both cameras can be switched simultaneously for continuous assessment of the zone associated with these cameras. The breakdown of the perimeter into zones can be accomplished for a CCTV system in conjunction with an intrusion detection system, such as microwave or E-field. In this case, every single or dual

Table 2 Computation for Method C

		Side	*AB*	*BC*	*CD*	*DE*	*EF*	*FG*	*GH*	*HK*	*AK*	*Total*
Method	*Cameras*	*Length (ft)*	*500*	*600*	*300*	*1,000*	*1,000*	*700*	*450*	*400*	*500*	*5,450*
C	Cameras with 75-mm lens		2	—		4	4	2	2	2	2	18
(Fig. 5)	Cameras with 50-mm lens		—	4	2	—	—	2	—	—	—	8
	Towers		1	2	1	2	2	2	1	1	1	13

intrusion detection zone will have two opposed cameras which will be called up automatically by intrusion alarm. No activation of CCTV cameras from an adjacent zone is required, and there will be no CCTV perimeter cameras associated with more than one intrusion detection zone, permitting dedication of each pair of cameras to its intrusion zone.

Methods A and B do not have this provision and have multizone assignment for each CCTV camera, creating confusion for an operator and requiring continuous operator attention to determine which area is under assessment.

COST EFFECTIVENESS

Method C is the most cost-effective method for CCTV camera layout. The comparison of equipment quantities is shown in Table 3. Method B is the basic method, and the quantities for this method have been assigned on a 100 percent basis. The percentages for Methods A and C are determined in comparison with Method B. Utilizing the opposed-camera concept, the estimated savings are as follows:

Equipment and Materials

Camera with associated accessories	35%
Towers	50%
Cables	25%

Installation

Labor	40%
Materials (conduits, j-boxes, foundations for tower, etc.)	30%

Maintenance

Labor	30%
Spare parts	25%

Table 3 Comparison of Equipment Requirements

Equipment	*Method*	*Quantity*	*Percent*
	A	57	139
Cameras	B	41	100
	C	26	63.4
	A	57	196.5
Towers	B	29	100
	C	13	44.8

CONCLUSIONS

Utilization of the opposed-camera concept for CCTV camera layout for a power plant perimeter has the following advantages:

1. Reduction of installation and maintenance costs from 25 to 35 percent.
2. Provision of better interface with the intrusion detection system by dedicated CCTV and intrusion detection perimeter zones.
3. Pertaining to the human factor—Provision of a logical sequence or order in video switching by zones, thereby reducing security operator tension.

REFERENCES

1. "Physical Protection of Plants and Materials," Title 10, Code of Federal Regulations, Part 73, U.S. Government Printing Office, Washington, DC.
2. "Acceptance Criteria for the Evaluation of Nuclear Power Reactor Security Plant," NUREG-0908, U.S. Nuclear Regulatory Commission, Office of Nuclear Material Safety and Safeguards, Washington, DC.
3. "Basic Considerations for Assembling a Closed-Circuit Television System," NUREG-0178, U.S. Nuclear Regulatory Commission, Office of Standards Development, Washington, DC.

An Evaluation of the Use of Low Light Level Television for Intrusion Control Along an International Boundary

Harry D. Frankel
Stephen Riter

Abstract. The U.S. Immigration and Naturalization Service has recently installed a test surveillance system using low light level television for operation by the Border Patrol in the El Paso, Texas area. The system has the potential not only to gather data on the effectiveness of the equipment used but to serve as a test bed for the evaluation of other systems.

The system is made up of nine low light level TV cameras, one very low light level TV camera, and one CCD camera. These cameras have been positioned so as to assure 24-hour video coverage of the most commonly used routes of entry along a nine-mile segment of the El Paso/Juarez international border. The cameras can be manually controlled or programmed to automatically sweep an area. In addition, there is a motion detector and the capability for on-line insertion of operational data. Also the system automatically records the location and time of key events and generates a continuous log of system activity.

Preliminary test and evaluation data has demonstrated a significant increase in field unit effectiveness with the system. There is also a potential for additional improvement beyond that which can be achieved with current manning levels.

BACKGROUND

The Border Patrol as the mobile uniformed enforcement arm of the United States Immigration and Naturalization Service (INS) is responsible for detecting and pre-

1985 Carnahan Conference on Security Technology, University of Kentucky, Lexington, Kentucky, May 15–17, 1985.

venting the illegal entry and smuggling of aliens into the United States. Border Patrol agents perform their duties along and in the vicinity of the 6,000 miles of international boundary and the Gulf Coast utilizing motor vehicles, boats, aircraft, horses and foot patrols. One of the Border Patrol's major methods of carrying out this responsibility is the performance of the linewatch function, the observation and patrol of the southern, northern and coastal U.S. borders to prevent intrusion.

It is believed that nearly 90% of all illegal entries occur along the southern border with Mexico. The entire area within 25 miles of the border is patrolled by agents with law enforcement authority. While the most likely routes of entry are known, it is very difficult for this force to control the nearly 1,970 miles of this border. Obviously there are enormous areas which cannot be continuously observed. Closed circuit television when properly sited and effectively managed has the potential to increase substantially the observation of these areas. This is particularly true when the border is adjacent to or cuts through urban areas. Because urban areas are easily approached from the Mexican side by transportation systems and provide a home for many people wishing to illegally enter, they are believed to account for a considerable percentage of the illegal entries. Also, once the alien enters a U.S. urban area there are a variety of opportunities to hide, blend into the local population or obtain transportation to the interior.

In order to determine the effectiveness of a closed circuit television system in support of the linewatch function, INS has recently completed the installation in El Paso, Texas of an initial configuration of a system of television imaging devices and means for their remote control. For the past six months the system has been operated in a test mode. The objectives of these tests are to:

- Prove through formal testing the improvements in linewatch which an initial configuration and subsequent variations can provide in El Paso.
- Determine standards and guidelines for selecting and integrating imaging systems and controls for support of linewatch in any Border Patrol sector.
- Explore new and evolving technologies as they may contribute to reduction of manpower loads, improvement of linewatch effectiveness, improvement of safety and security, and reduction of linewatch costs.

In this paper we describe the initial system configuration and report on the test results obtained.

SYSTEM DESCRIPTION

The initial configuration of the system consists of 11 low light level TV cameras placed at nine locations along a nine mile segment of the border. Figure 1 shows the approximate geographic location of the sites in relationship to the border. It should be noted that the border which is defined by the Rio Grande River separates the city of El Paso, Texas with a population of approximately 500,000 from Ciudad Juarez, Mexico, a city with a population estimated to be 700,000. The border is within walking

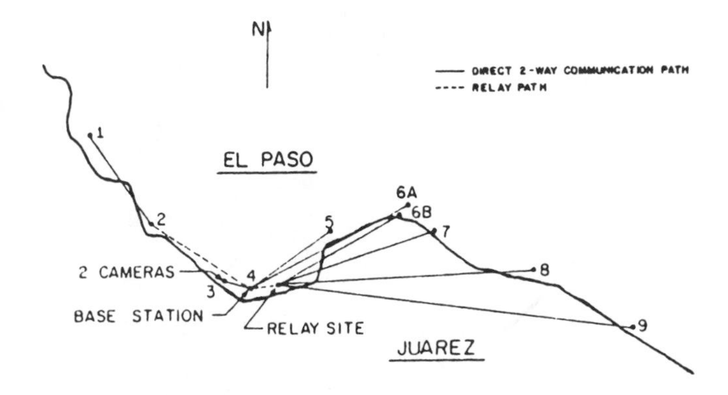

Figure 1. Initial camera deployment and communication network.

distance of the downtown area of El Paso and close to residential areas in both Mexico and the United States.

The camera sites were chosen and the cameras positioned so as to assure 24 hour continuous coverage of the most commonly used routes of entry. They are linked by a microwave system to a control center located at Site 4 on the figure. Because of poor line-of-sight conditions, paths between some of the cameras and the control center are relayed via a repeater. All cameras are located on 60 foot poles with the exception of the two cameras at Site 3, the camera at Site 6A, and the one at Site 6B. Ten of the cameras have remote azimuth and elevation control, variable fields of view, and zoom lens capable of going from 16 mm to 160 mm focal lengths.

The camera at Site 1 is a very low light level TV camera capable of seeing under starlight conditions. The other cameras with the exception of the camera at 6B, are low light level TV cameras capable of operating under quarter moon illumination. A fixed focus CCD camera is located at Site 6B. The mix of equipment was chosen to meet the operational requirements presented in El Paso, be a reasonable representation of conditions likely to be encountered elsewhere and to provide a vehicle for comparing different equipment under varying operating conditions.

The heart of the system is the base station located at Site 4. From the station Agent-Operators control the cameras, observe the video images, and dispatch field units. A schematic showing the initial base station configuration along with its connection to a typical remote camera location is presented in Figure 2. The microwave link is used to transmit video images from the camera to the base station. A separate control channel is used for sending control signals from the base station to the camera. Each video signal is assigned a separate microwave channel so there is no possibility of video interference. However, the control signals are all sent on a single microwave channel with address signals preceding each control signal.

In Figure 3 we show a representation of the base station control equipment. With it the Agent-Operators can simultaneously observe the video images from the eleven TV cameras on eleven 9″ monitors. The cameras can be programmed to automatically sweep an area or be fixed on a single location. The Agent-Operators can call up on

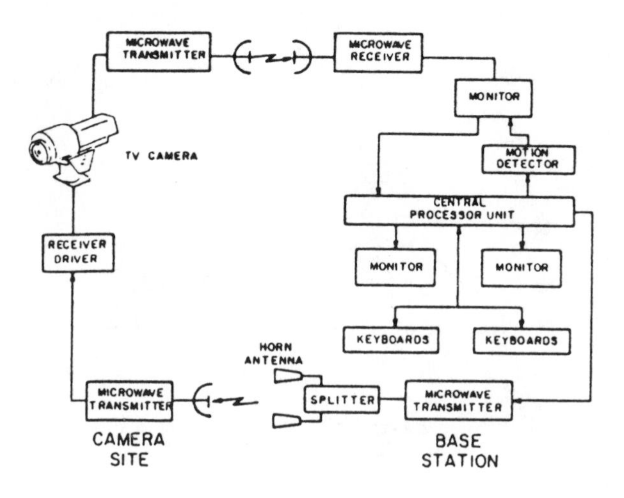

Figure 2. System configuration.

either of two 12″ TV monitors images they wish to observe more carefully. They can also override the automatic control of the camera and manually control its focus, pan and tilt. In addition, the system has a motion detector which automatically alerts the Agent-Operators to activities in designated area of each camera's field of view. A special sensing system with synthetic voice is used to alert the Agent-Operators to the approach of people to the camera locations. There is also voice communications with Border Patrol units in the field.

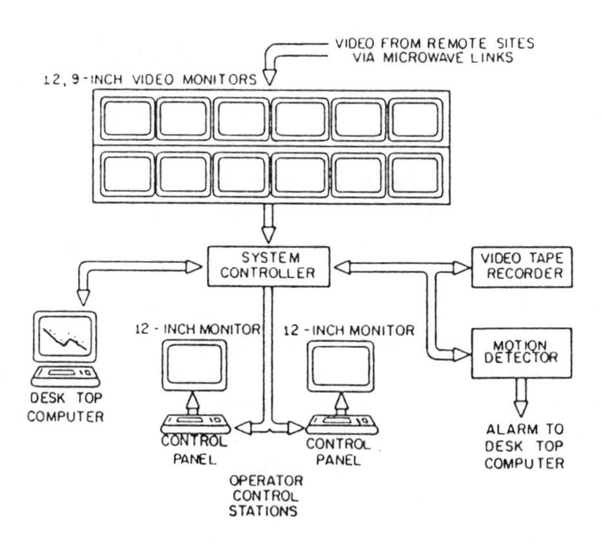

Figure 3. Base stations.

The programmable control mentioned above enables the Operators to select for each camera ten pointing directions and focal lengths by entering an appropriate address. The cameras can be assigned to sweep between these locations with preassigned dwells at each location.

The system also includes a video tape recorder for collecting evidentiary data and a data management computer to aid in the collection of management information as well as test data.

When an Agent-Operator begins a shift, the data management computer leads him through a series of data entries which require him to record environmental conditions, shift number, time and date, the status of all cameras and related equipment, and other operational data. During the course of the shift, the computer aids Agent-Operators in collecting data on people observed, number of men, women and children sited, number of units dispatched and number of people apprehended. This information is collected as a function of camera location and time of day.

DATA COLLECTION

During the preliminary operation of the test bed three classes of data were collected: coverage estimates, shift activity data, and shift operational data.

Coverage estimates are independent estimates obtained from experienced Border Patrol agents using modified Delphi techniques of the percentage of intrusions observed with each camera which would have gone undetected without the system. These estimates, which are by nature subjective, were used to provide baseline data from which improvements in system performance could be judged based upon the actual number of people observed and apprehended during the course of the system test. Shift activity data consisted of descriptions of critical events per shift observed by the Agent-Operators. This information was collected on the data management computer and included the number of intruders detected by camera, assignment or reason for nonassignment of field units, classification of intruders detected, time of detection, time of apprehension. Shift operational data described the conditions under which the operational data was obtained. This includes environmental data, number of agents on duty and status of equipment.

EVALUATION OF DATA

The system was operated in a test mode for a three-month period beginning September 15, 1984. The use of Delphi techniques to estimate the improvement suggests that at least 75% of the individuals detected with the system would have gone undetected using conventional linewatch procedures. Of the people observed 90% would have been apprehended if sufficient manpower were available. It should be noted that positions have been authorized and personnel are being trained to bring the Border Patrol to a level where this improvement can be realized.

A number of other factors were tested, including the number of operators required to operate the system and the number of cameras a single operator could carefully monitor.

One of the most critical aspects in evaluating a system such as this is the interface between the operator and the system. Detailed interviews were conducted with the Agent-Operators. They were asked a variety of questions designed to elicit their opinions of system effectiveness and provide ideas for long-term system improvement. In general, the operators felt that the system contributed greatly to organizational effectiveness. Most Agents said upon detailed questioning that it was difficult for them to scan more than 4 monitors at once. If activity was observed it was not possible for them to watch the other monitors.

The system contained a motion detector but the motion detector was seldom used. It had high false alarm rates and consequently the operators felt that it had negligible value. When given an opportunity the Agent-Operators would turn it off.

We also questioned the Agent-Operators about the value of being able to automatically sweep an area. They found it even more difficult to concentrate on the monitors when the cameras were in the sweep mode. When given the opportunity they would turn this feature off.

CONCLUSIONS

It is too early to conclusively answer the fundamental question of how much a system like this can improve effectiveness. However, there are already unmistakeable trends which are of importance to INS and which should be noted by others involved in the design of intrusion control systems.

First and most importantly, a closed circuit television system provides considerable improvement in detection capability. As mentioned, our studies indicate that over 75% of the people observed would have gone undetected using current methods. As importantly, by carefully observing the illegal entries, one can pick and choose the people against whom to commit scarce manpower resources. Regretfully, without sufficient manpower it is not possible to realize a significant increase in apprehensions.

Closed circuit television systems are complex technical systems requiring trained maintenance personnel. We experienced difficulty in keeping all cameras operating through the initial period. At one time as few as seven were functioning and we averaged nine over the period. Part of the problem may have been due to the usual difficulties involved in bringing a new system on line. We believe that a system like this requires the assignment of at least one full-time technician.

In designing a system such as this, one must pay as much attention to human factors consideration as any single technical factor. We learned that the Agent-Operators can be depended upon to do a good job of monitoring television images on a continuous basis but that one must provide them with an appropriate environment in which to do this. The environment should feel open, but be isolated from distractions. The monitors need to be placed so as to minimize eye strain and physical discomfort. We were disappointed to discover that the number of monitors one person could carefully concentrate on was so few and that the motion detectors and automated sweep capability were considered to be of limited value. As pointed out earlier when given the opportunity the Agent-Operators turned off both of these features. We remain convinced of their value and have initiated an activity intended to extend the capabilities of conventional motion detectors working in conjunction with an automatic sweep.

Finally, such systems generate a wealth of management information. However, techniques to fully utilize this information need to be developed and managers convinced of their value.

In conclusion, we have demonstrated the value of this system as a test bed for evaluating methods of intrusion control. In the future we hope to integrate infrared units into the television system, develop automated techniques which might reduce or eliminate the number of operators required and test technologies and techniques developed by other organizations.

Enhancement of Direct Visual and CCTV Surveillance/ Assessment Functions at the Perimeters of Controlled Access Sites

Francis P. Pfleckl

Abstract. Illumination for the direct visual and CCTV surveillance/assessment functions at the perimeters of controlled access sites is costly and, for viewing ranges in excess of 600 feet, it is usually inadequate, especially during inclement weather such as fog, heavy rain, or falling snow. This same illumination is also a major contributor to the unique, night time signature of the site. Augmenting an existing lighting system with high-intensity reflective material, strategically placed, has been found to improve upon several expensive and inadequate conditions. Through the application of highly reflective material and luminaires collocated with the mode of observing a scene—either direct visual or CCTV—it can be shown that: 1) the amount of primary power is significantly reduced, 2) the probability of correct assessment at far ranges is improved, 3) the site signature from security illumination is lessened, 4) some poor weather surveillance and assessment capabilities are increased, 5) the cost-effectiveness ratio is improved. The end result is that the capabilities of security guards, site vulnerabilities, and the cost to effectiveness ratios are benefited.

INTRODUCTION

Among other things, the probability of detecting intruders can be improved if an effective visual environment is designed and implemented. In the past, attempts have been made to utilize artificial surfaces for reducing maintenance costs accountable to

1983 Conference on Crime Countermeasures and Security, University of Kentucky, Lexington, Kentucky, May 11–13, 1983

the perimeter clear zones of controlled access facilities, while at the same time the visual environment is improved. Previous efforts to determine the optimum artificial surfacing technique for perimeter clear zone ground covers have been inadequate.

Additionally, security lighting is emerging as a complex system involving more than placing luminaires on poles around the perimeter of an installation. Recent studies have indicated that special lighting designs have the potential to increase the visual response of security guards. Furthermore, it has been observed that certain lighting techniques affect the psychological, physiological and behavioral responses of intruders, security personnel and response force personnel.

The increasing cost of energy has also demanded an evaluation of existing lighting systems. With the advent of Closed Circuit Television (CCTV) as an alternative to direct visual surveillance and assessment of controlled access areas, much work has been done in developing new concepts, new designs, and new hardware for security lighting systems to permit expanded use of CCTV.

The state of the art in security lighting and artificial ground covers is continually crossing new thresholds. Most literature that is available today gives recommendations for industrial, residential, or highway lighting systems. Little information, if any, is available which attempts to correlate illumination levels with ground cover combinations to parameters dealing with security force responses. Where literature on this subject does exist, it primarily deals with the performance of visual tasks under controlled conditions.

The following article will address the response of security guards to random intrusion attempts during "controlled" field conditions under different lighting and ground cover combinations. Parameters such as lighting levels, reflectivity, and the probability to correctly assess a target at distances of 600 feet or greater will be discussed.

PROBLEM DESCRIPTION

A part of the operating expense in security is maintaining an adequate level of illumination, particularly in perimeter security lighting.

With any perimeter lighting system, the illumination that is used to maintain high security is also the same illumination which pinpoints the location of the facility, especially when the facility is observed from the air. At times, it is better not to advertise so "brightly" where a protected area is located. Herein lies the problem—how to maintain an adequate visual environment while reducing the site signature and controlling utility costs accountable to perimeter illumination. Also, suppose that the primary source of power for lighting is incapacitated along with the UPS system: in this scenario, can an adequate backup security lighting system be readily available so that accurate assessments of the clear zones can be maintained? The Lighting and Deterrent Branch of the Counter Surveillance/Counter Intrusion Laboratory, US Army MERADCOM, Fort Belvoir, VA attempted to answer these questions through a contract which was awarded to Mission Research Corporation in February 1981.

Definitions

Before proceeding, it is appropriate to define two technologies that are fundamental to this discussion. These technologies are Reflector Technology and Assessment Technology.

Reflector Technology in physical security functions is the application of reflective materials for the enhancement of either the direct visual or the electro-optical (CCTV) environment of security forces in the performance of their surveillance and assessment functions.

Assessment Technology is the application of methods and materials to improve the capabilities of security forces to scrutinize protected zones, and recognize and identify targets in a timely manner.

TYPICAL PERIMETER SECURITY CONCERNS

Several combinations of visually aided assessment systems are being used today. The majority of these systems use guards in towers for detecting intruders at a perimeter. Some installations have a single barrier around the perimeter; usually this barrier is a chain link fence. Most sensitive military sites use a double chain link fence to delineate their boundries. The primary function of the guard is alarm assessment. To assist the guard in assessing a target or an alarm, several sites rely upon the combination of Closed Circuit Television (CCTV) and tower guards. CCTV is used as a cost-saving factor—CCTV can be placed to cover areas which are not visible to the guard from his stationary location.

Other sites use Intrusion Detection Systems (IDS) for detection and CCTV for alarm assessment. There are no towers in this case, only roving patrols.

In all of these cases, lighting is paramount in importance. Each case needs a specific lighting system for optimum performance. The lighting system must be designed to support direct visual surveillance or CCTV surveillance, or both. What is designed for one perimeter is not necessarily the optimum for others. Trade-offs are necessary. In addition to supporting detection and assessment functions, the lighting system must also support a response force approaching the point of alarm. The response force must not be unnecessarily illuminated or hampered by glare. Glare in the eyes of an intruder is beneficial but can not be tolerated in the eyes of the security forces.

In summary, for military applications in particular, the lighting system must support the functions of surveillance, assessment and interdiction, as well as being economical. At the same time, it should limit, as much as possible, the definition of site geometry and topography to the outside world. In other words, site signatures should be reduced while security is maintained.

DETECTION AND RECOGNITION AIDS

Contrast and motion are two important principles in the detection and recognition of a target. For purposes of this article, contrast will be understood as the difference in brightness (luminance) between a target and its background divided by the luminance of the background. In accordance with this definition, contrast can be either positive (a bright target) or negative (a dark target). Increasing the contrast results in increased detection if target speed is held constant. If contrast is held constant, an increase in speed increases detection.[1] Size of a target and its duration of exposure also affects its detectability; however, this detectability varies with contrast and background luminance.

When contrast is positive, the luminance of the target is greater than the background. The objective of present lighting systems is to cause positive contrast. Since the reflectance of the target can not be controlled by the design of the lighting system, the reflectance of the background usually is reduced to increase contrast. Materials used in clear zones, such as backgrounds between security fences, are usually dirt, grass, or asphalt. Typically, these materials have a reflection of 50% or less. Since it can not be assumed that an intruder would be cooperative enough to wear highly reflective materials, positive contrast has the following advantages: a) there is no theoretical limit to the amount of incident light that can be used; b) with the contrast positive, contours of a target can be readily seen; therefore, this provides valuable information to identify a target; c) it has been shown that a light target on a dark background is more easily seen than a dark target on a non-dark background. However, increasing the level of illumination means increased utility bills and accenting the site signature.

Negative contrast, on the other hand, exists when the target reflectance is less than the background reflectance. In this case, a target looks dark against a bright background. Even though obtaining a bright background may seem to be ludicrous, there are some feasible methods to satisfy this need. Negative contrast concepts place more importance upon the effectiveness of the background than the positive contrast concept. One concept to assure a bright background is to use an active light source for the background. Because the length of a zone under surveillance is usually large when compared to the height of the point of observation (the guard or TV camera is usually in a tower or on a pedestal), separate discrete sources would be required. Implementing this concept does not have a high practical value.

Another method to make a bright background is to use components with large optical gains. Examples of these components are glass beaded surfaces (spherical reflectors) and cube corner reflectors. The beads in beaded surfaces act as small lenses which concentrate incident light on a back surface—either the back of the bead itself or some other reflective surface. The light is internally reflected and returned toward the illuminating source. Examples are traffic signs, movie screens and beaded roadway lines.

Cube corner reflectors are tetrahedrons. A light ray entering the tetrahedron is reflected from three sides of the cube and the ray exits the cube in a path parallel to the entry path. Internal reflection is nearly total and, therefore, it is a highly efficient reflector. Examples of cube corner reflectors are automobile and bicycle reflectors.

The above components can be lumped into a category labeled "retroreflectors" because the source of illumination is reflected back toward the source. Diffuse reflection results when the incident ray is scattered in all directions. Mirror reflection reflects the incident light at an angle equal to but in an opposite direction to the incident ray.

A third method of achieving a bright background is to use ultra violet illumination (within safe limits) and retroconverters. Retroconverters use a lens to focus Ultraviolet (UV) illumination on a fluorescent screen. Visible light from the florescing screen is collimated back in a parallel path to the incident UV illumination. The use of UV luminaires and retroconverters may be advantageous in inclement weather because there would be low losses due to veiling or scattering of the incident UV ray. Using this particular concept has shown some promise in improving surveillance and assessment functions during inclement weather, and it may be proven to be more effective than other visual systems used today. It would definitely be less expensive than an IR system or a millimeter wave length system.[2]

The spherical retroreflectors can be commercially purchased in sheets of different sizes and lengths, and because of this, they have a wide application augmenting security lighting systems. Cube corner reflectors are available in small diameter circles or thin short strips; these, therefore, have limited use in augmenting security lighting systems. Retroconverters are not readily available at this time to make them practical for security lighting systems. The remaining discussion will concern only spherical retroreflectors.

RETROREFLECTOR CONFIGURATIONS

Commercially available retroreflector material is made of microscropic spherical glass beads embedded in a durable transparent plastic. The sheeting has a reflective coating behind the glass bead layer, and it is manufactured in sheets which can be cut to various widths. Different methods are available for mounting the material. Some material is available with an adhesive backing layer, and others must be heat treated for proper adhesion. The plastic is easily cleanable and very conformable. The surface upon which the sheeting is mounted must be clean, dry, smooth, rigid, non-porous and weather resistant (Figure 1).

Exclusion areas at controlled access sites for US Army bases are surrounded by a double, seven-foot-high chain link fence. A clear zone of approximately thirty-five feet exists between the fences. It is between the fences that a retroreflective array can be placed. Also, the retroreflective material can be attached to the fence fabric and

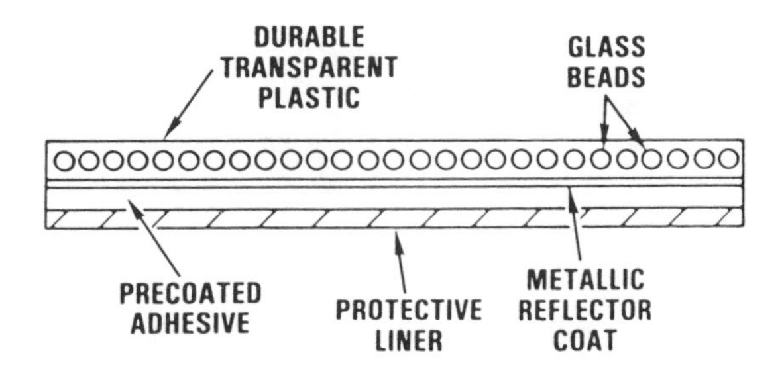

Figure 1. 3M reflective sheeting.

placed at other strategic locations around the perimeter and within the compound (such as on the doors of storage bunkers) in order to afford roving patrols and response forces the efficacious use of the optical principles involved.

In the clear zone between fences, a retroreflector array can be placed in strips to span the entire thirty-five foot length between the fences. The strips should not be installed at a height greater than six inches.

If the height of the strips is greater than six inches, it may be possible for an intruder to hide behind an individual strip and low-crawl across the clear zone without visual detection from tower or stationary guards. Rain will not greatly affect the optical gain of the retroreflector unless the water is deep enough to cover the array. Any snow that obscures the arrays will defeat their purpose; all is not lost, however, because anyone traversing the snow-covered area will leave a trail which can be assessed. Obscuration of the array by leaves, sand, or growing grass does limit its effectiveness; some maintenance is required in this regard. If flexible mounting hardware is used to hold the retroreflector array, then mowing will not be a major difficulty. The concept is that after a pass of the lawn mower the apparatus will spring back into position (Figure 2).

Geometry Makes It Work

Around the perimeter of most security sites are High Pressure Sodium (HPS) luminaires. In particular, 250 watt, 120 VAC luminaires are popular in the U.S. These luminaires are placed within the compound, along the inside fence. This shall be considered as the "standard" or "broadside" lighting system. For broadside lighting, the sight line from the tower guard to the target is nearly parallel to the fence line, and the illumination is nearly perpendicular to the sight line. The illumination is directed

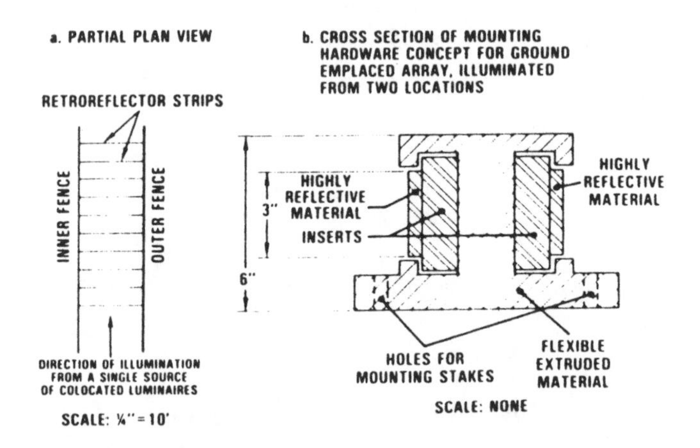

Figure 2. Configuration of the retroreflector array.

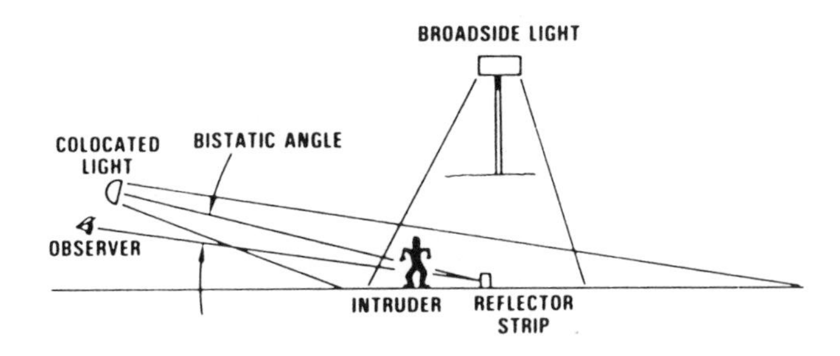

Figure 3. Geometrics for collocated and broadside lighting concepts, and reflector arrays.

downward into the clear zone from the luminaires which are mounted about thirty feet overhead on poles equally spaced around the perimeter.

A collocated light is one located near the observer. Effectiveness of a collocated light source when used with retroreflectors is strongly dependent upon the bistatic angle. This angle lies in a plane containing the source, the retroreflector, and the observer. A bistatic angle is an angle between the rays from the reflector to the light source and from the reflector to the observer. In general, bistatic angles must be small, less than 1° for maximum effectiveness. It is to be noted, however, that enhanced capabilities are obtainable even approaching bistatic angles of 30° (these configurations are shown in Figure 3).

The combinations of the collocated luminaires with the high-intensity retroreflective arrays present a target dramatically in negative contrast—the target appears as a dark silhouette against a bright background (Figure 3).

ASSESSMENT TECHNOLOGY

The concept of using an artificial ground cover to enhance a visual environment is not new. What has not been investigated and reported before is the use of a high-intensity retroreflective array and collocated luminaires in perimeter security operations. This concept is plausible but to evince the concept, tests under actual field conditions were necessary. These tests were conducted at Range 1 of the MERADCOM Test Area, Fort Belvoir, VA.

Range 1 is a ten-acre area of grass-covered sloping terrain with two parallel, seven-foot-high chain link fences. The chain link fences run for 900 feet. A clear zone of thirty-five feet separates the parallel fences. A clear zone of thirty feet or more exists on the outside of each fence. Wooden mounting poles for luminaires, electrical power and a guard tower are available at the site (Figure 4).

Real-time field tests were conducted at Range 1 by Mission Research Corporation, Santa Barbara, CA under Government contract to determine the applicability, and the advantages and disadvantages of the new concepts.

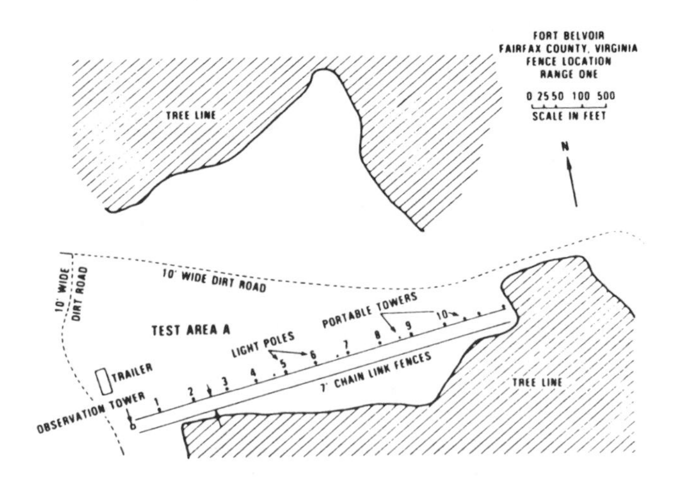

Figure 4. Lighting and barriers test site.

The reflective array consisted of high-intensity-grade sheeting, manufactured by 3M Company. The three-inch-wide, Series 580 material was used. Also, three 12 VDC, 37½ Watt (W), instant-on tungsten iodide lamps were used as the collocated luminaires mounted at the guard tower, at approximately thirty feet above the finished grade and 900 feet from the extreme end of the double chain link fence. The array was placed at the finished grade between the fences in twenty-one sections evenly spaced over a range of 700 to 900 feet away from the base of the guard tower.

As a means for comparing and evaluating the results of the tests of the new concepts, a "standard" broadside lighting system was also used in a separate series of tests. This "standard" broadside lighting system used four, 250 W, 120 VAC, HPS luminaires to illuminate the area of concern.

Night-time tests were conducted during December, 1982 using contractor personnel to act as guards and intruders. Two modes of approach were used for intrusion attempts; they were the run and low-crawl. Distances of 700, 800, and 900 feet away from the guard tower were used for varying the range of the intrusion attempt. Several "false alarms" were included in the actual attempts as a means to protect against cueing the guards; other controls were also used to limit the possibilities of cueing the guards. "False alarms" were artificially generated by causing an intrusion detection system to annunciate when an intruder was not attempting to penetrate the site.

The responses of the tower guards and CCTV monitoring guards were noted and recorded. The guards were required to state whether there was an intrusion attempt and, if there was one in progress, the guard was to report whether it was a run or a crawl, and at what range the intrusion attempt was taking place. The time to detect was also noted. All of these tests were conducted with the different lighting and retroreflector array combinations.

From previous studies, it was learned that the probability of detection decreases at distances in excess of 600 feet from the point of observation; all tests for the above investigation were conducted at ranges in excess of 600 feet so that meaningful statistical results could be obtained and judgments made about the benefits and deficiencies of the concepts tested.

The CCTV camera used during the test was a COHU 2850C series, automatic iris, low light level, self-contained environmental camera. It was placed approximately 200 feet in front of the beginning of the reflector array. The camera responded well to the IR rich luminance. So well, in fact, that the probabilities of correct assessment were high and not statistically significant for comparative analysis.

Even though the camera in these tests was useable, other cameras that have a silicon-intensified target, automatic iris, and a second intensifier may not perform as well as needed while the retroreflectors are used. This is because the camera's amplifiers apparently saturate and no useable picture is transmitted to the monitors.

LIFE CYCLE COSTS

No investigation would be complete without a brief discussion of the costs involved in implementing the new technology.[2] The following is such an attempt.

Retroreflectors

It is assumed that the retroreflectors would be mounted in a clear zone requiring periodic mowing. The level of the grass should be kept below the bottom of the reflective sheeting to retain maximum brightness. This requirement may impose a need for increased frequency of mowing in climates and soils favoring grass growth.

Dust or mud on the plastic face of the reflective sheeting will degrade performance. Cleaning agents such as naptha can be used and such agents are needed for installations adjacent to roadways where oily deposits are expected. Plain water should be sufficient in isolated regions. It is estimated that sixteen to twenty man-hours would be required to clean the retroreflectors in a 100 m zone. The frequency of cleaning is a total unknown. Rainfall may be sufficient to keep the retroreflector faces clean, but rain may also splash mud on the face, requiring more cleaning.

With the retroreflectors emplaced, there should be no impact upon maintenance of line sensors buried in the same zone. However, they affect performance of microwave or Infrared sensors. Presence of the aluminum strips in the zone may degrade performance of both of these sensors.

The lighting configuration assembled for testing used three 37.5 watt tungsten iodide lamps, each providing 70,000 candlepower. The combination produced an illuminance of approximately 0.2 fc on a surface perpendicular to the line of sight at a distance of 900 feet (274 m) from the lamps. Results presented show that, with this level of illumination, guard performance is significantly better than with the broadside lighting configuration providing the same illuminance from polemounted HPS 250 luminaries. No work has been done to determine how much the illuminance using retroreflectors could be reduced without degradation of performance. The major impact on the required light level is the number of lamps involved and the 300-hour design life of the lamps. The replacement cost for the relatively inexpensive tungsten iodide lamps dominates costs over 10 years. Discussions with manufacturers indicate that lifetime of the bulbs can be increased 10 to 20 percent by operating at reduced voltage. Further work leading to optimum use of tungsten iodide lamps is needed. When the life of the tungsten iodide lamps increases, the relative life cycle costs of the concept would be much improved.

At the end of 10 years, the retroreflective sheeting would be replaced. It may be possible to install new sheeting over the old sheeting. If the old sheeting has to be removed the aluminum would have to be cleaned before the new sheeting is applied. A stripper is available from 3M.

The cost differentials emphasize the need for an illuminator with a longer lifetime. (Assuming an HPS 250 in a POWR-SPOT luminaire, the 10 year O&M is about $10,000.) Use of tungsten iodide lamps and retroreflectors, in the configuration used for testing, would be justified on the basis of improved guard performance, rather than cost reduction.

High-Pressure Sodium Luminaries

It is assumed that 250 watt High-Pressure Sodium lights are used. It is further assumed that they are mounted on 35-foot steel poles with 6-foot arms and concrete bases. Two poles are needed for a 100 m zone. Buried armored cable is assumed for power distribution. Since there is no reasonable way to estimate a distance from the transformer to the perimeter zone, cost for buried cable running the length of the zone is used. This seems reasonable for comparison purposes if it is considered that prime power is furnished at the zone for both HPS and tungsten iodide lights.

Lifetime for HPS lights is given as 15,000 to 20,000 hours. Lifetime used here is 4 years (17,600 hours).

Costing does not include CCTV. Although crushed rock coated with paint containing reflective beads was not tested, data are available for costing so costs for a configuration of tunsten iodide lights/beaded crushed rock is included. Table 1 gives

Table 1 Relative 10-Year Cost

Lighting/Ground Cover	*Ratio of Costs*
HPS 250/Crushed Rock	1
HPS 250/Grass	0.97
Tungsten Iodide/Retroreflectors	1.15
Tungsten Iodide/Beaded Rock	2.39

the lighting/ground cover configurations and ratio of 10-year cost of standard HPS 250/Grass.

CONCLUSIONS

The empirical results from the tests just described show that the reflector concept has great potential for improving the effectiveness of perimeter security. The advantages that were determined are as follows:

1. Lighting levels were reduced by a factor of ten. With the reflector array in place, four 250 Watt, High-Pressure Sodium (HPS) luminaires were replaced with three 37½ Watt tungsten iodide luminaires with good results.
2. The probability of correct assessment was either maintained or improved during the combined use of the three collocated luminaires and the reflector array.
3. Typically it takes six to eight minutes for the HPS lamps to reach full brilliance from a cold start. The tungsten iodide luminaires, however, are instant-on, even from a cold start.
4. If the power source or cable is cut for the primary perimeter illuminators, the collocated luminaires can be on line by simply connecting them to a vehicle battery.
5. The ability to reduce lighting levels also reduces the site signature.
6. By placing retroreflective material in strategic locations the effectiveness of response forces in their fighting positions is increased.
7. Portable luminaires, such as flashlights, can be used in conjunction with the retroreflective material to assure correct assessments of alarms by roving patrols.
8. It is believed that alarm assessment capability can be improved to some extent during inclement weather.
9. Background reflectance increased by a factor of forty.
10. Ten-year life cycle costs are comparable to the "standard" system.

The ability to effectively assess a target during reduced lighting levels has crossed a new threshold in physical security surveillance and assessment functions through the advent of reflector technology. The concepts are technically sound. They need to be proved acceptable and practical for users through further development and tests.

REFERENCES

1. H.E. Peterson and D.J. Dugas, "The Relative Importance of Contrast and Motion in Visual Detection," *Human Factors 14*, 1972.
2. R.C. Scott, T.J. Barrett, and R.W. Hendrick, Jr., "Lighting Systems and Artificial Ground Cover (U)," vols. 1, 2, and 3, Technical Report, May 1982, Mission Research Corporation, 735 State Street, P. O. Drawer 719, Santa Barbara, CA 93102. Prepared for Commander, US Army Mobility Equipment and Research Development Command, ATTN: DRDME-XI, Fort Belvoir, VA 22060.

3. J.H. Morgan II, R.J. Ely, L.E. Jones, L.B. Scheber III, M.A. Smith, and M.B. Upson, ''A Quantified Approach to Perimeter Barriers and Lighting Development (U) Final Report,'' April 1981, The BDM Corporation, 7915 Jones Branch Drive, McLean, VA 22102. Prepared for Commander, US Army Mobility Equipment Research and Development Command, Fort Belvoir, VA 22060.
4. *IES Lighting Handbook*, 5th ed., 1972, Illuminating Engineering Society, 345 East 47th Street, New York, NY 10017.

PART V

Security System Applications of Fiber Optics

The unique attributes of fiber-optic technology make it ideal for security system application. To name two such applications, it is immune to electro-magnetic interference and more importantly it is virtually impossible to tap into for information or to violate without detection. Important to some applications is that it is safe in explosive environments.

Fiber-optic technology applications in security are in three general areas: secure voice and data communication, perhaps the most important in these times of extensive espionage; communication links between security sensors and controls and alarm annouciators; and excellent potential in actual sensor design.

Since these articles were published, fiber-optic technology has become widely available in the first two applications. A number of companies commercially manufacture fiber-optic systems for general communication and as links in electronic intrusion detection systems. In addition to the use in intrusion detection systems, fiber optics are now being employed in CCTV systems as links between cameras, monitors and recorders. The cost of these fiber-optic links is now competitive with the old hard wire links. In some applications, fiber-optic systems are cheaper.

It is predicted that the fiber-optic medium will be readily applied to a vast array of sensor designs. Presently sensor designs which benefit from beam bending and fiber breakage are commercially available. More sophisticated designs will soon be commercially available.

The two articles in Part V are not only highly informative about fiber-optic technology individually, but they are also highly complementary. The Rarick article presents an excellent explanation of fiber optics along with the historical development and a view to the future. The Güttinger and Pfister article illustrates many advantages of fiber optics and provides more specific discussion in many areas, especially the potential of fiber optics in sensor design.

A Short Review of Fiber-Optics Technology

Jay A. Rarick

Abstract. The purpose of this review shall be to discuss the basics of fiber optics in a general way, with attention being shown to the development of fiber optic technology over the last fifteen years.

Included shall be a short description of the past use of lightwave communications, of fiber optic technology from the containment of light within an optical waveguide to a discussion of the types of fibers in use today, connectors, splices, emitters and detectors, fiber bandwidth and optical multiplexing, with a word to the advantages and disadvantages of fibers as a communications medium. Also included shall be a review of the historical development of optical fiber communications from its theoretical beginnings to the invention of the laser up to the present time. This shall include the principle innovators, and a review of the current state of the art in optical fiber technology with the current record performance systems and a look beyond to what the future holds.

INTRODUCTION

For centuries, man has sought to communicate with other men at a distance. He has been increasingly successful at this up through the age of electronics, but using light to carry information is an old concept which is undergoing a powerful resurgence.

Early examples of using light to carry intelligent information included the American Indians using smoke signals and the English building bonfires to warn of the Spanish Armada's approach. Closer to our modern meaning of information carrying is Claude Chappe's optical telegraph system of the 1790s which consisted of a system of semaphore stations on hilltops throughout France. "The system, which reputedly could transmit messages a distance of 200 kilometers in 15 minutes, remained in service until superceded by the electric telegraph."[1]

1984 Carnahan Conference on Security Technology, University of Kentucky, Lexington, Kentucky, May 16–18, 1984.

The first optical, voice-modulated channel was devised by Alexander Graham Bell in 1880.*

> In one system Bell focused a narrow beam of sunlight onto a thin mirror. When the sound waves of the human speech caused the mirror to vibrate, the amount of light energy transmitted to a selenium detector varied correspondingly. The light reaching the detector caused the resistance of the selenium, and thereby the intensity of the current in a telephone receiver, to vary, setting up speech waves at the receiver end.[2]

The invention of the laser sparked interest in communications engineers because of the laser's directivity and the high frequency of light which offered perhaps one thousand to ten thousand more times the information-carrying capacity of the shortest wave radio communications of the time.[3] "The quality of information that can be carried by a transmission channel . . . depends on the frequency of the signal, the higher the frequency, the greater the amount of information."[4]

In the course of this paper on fiber optics we shall endeavor to answer several questions. What was the innovation with fiber optics? What were the prior conditions to this innovation? What was the sequence of events and problems and interaction among the participants?

To address these questions, we shall investigate the phenomenology of fiber optics and then address the historical development of optical fiber technology. In conclusion we shall look at the state of the art and where fiber optic applications are going today.

FIBER-OPTIC COMMUNICATIONS

The Phenomenology of Fiber Optics**

An optical fiber, as is used today, is more correctly called a dielectric waveguide, for the propagation of electromagnetic radiation through an optical fiber is a guided wave phenomenon. The ground work for understanding these waveguides was laid down by James Clerk Maxwell in 1865. While we shall approach the workings of an optical fiber from a less mathematical level, it is important to remember that the propagation of the wave, as governed by Maxwell's equations, is a more correct description than that given by geometrical optics.***

An optical fiber, simply, is a fiber of very pure silica-glass with an outer layer or "cladding" of glass applied to it, thus forming the light guide. Light is contained

---

*Bell, A.G., "Selenium and the Photophone," Electrician, 5, (1880), p. 214.

**Unless otherwise stated the wavelength region of fibers and components in this paper are .80 μm – .9 μm.

***Several good review articles exist on the mathematical description of optical waveguides. Two are: Barnoski, Michael K., Editor, *Fundamentals of Optical Fiber Communications* (New York: Academic Press, Inc., 1976). Olshansky, R. "Propagation in Glass Optical Waveguides," Reviews of Modern Physics, Vol. 51, No. 2, p. 341ff, (1979).

within the fiber by means of a phenomena known as total internal reflection. Total internal reflection is described by a simple equation known as Snell's Law, which states that light moving across an interface from a region of higher refractive index to one of lower refractive index will be bent away from a line perpendicular to the surface of the interface where the light strikes. In mathematics this is stated:

$$n_1 \sin \theta_1 = n_2 \sin \theta_2$$

where n_1 and n_2 are the indexes of refraction for the two regions and θ_1 and θ_2 are the angles made by the incident and transmitted rays with the perpendicular (sometimes called a "normal"). This is shown clearly in the Snell's Law diagram (Figure 1). As θ_1 increases, θ_2 does also until θ_2 equals 90°. At this point, all the light of the transmitted ray just skims the interface and at any higher angle of θ, is totally confined to region one. This region, in the case of a fiber, is the core. The earliest demonstration of this was performed by John Tyndall for the Royal Society of London in 1870 where he illuminated a vessel of water and showed that where a stream was allowed to flow through a hole in the vessel, the light was conducted along the curved path of the stream.[5]

There are three basic types of fibers in use today. These are shown in Figure 2. In cross section they are a "step-index," "graded-index" or "single-mode" fiber. In the step-index the index of refraction is constant throughout the core. The result of this is a conceptually simple fiber in which light entering at one end travels different paths through the fiber depending on the angle at which it entered. Those entering at sharp angles are propagated at what are called high order modes and those which stay close to the axis are called low order modes. In a step-index fiber, high order modes

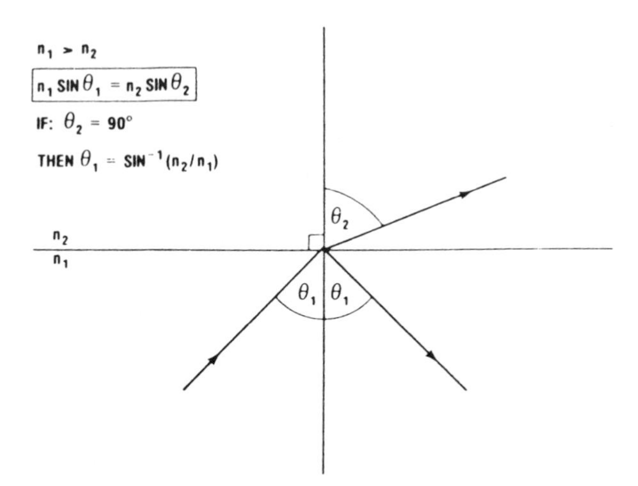

Figure 1. Snell's Law for refraction and reflection between two materials with indexes of refraction n_1 and n_2 respectively.

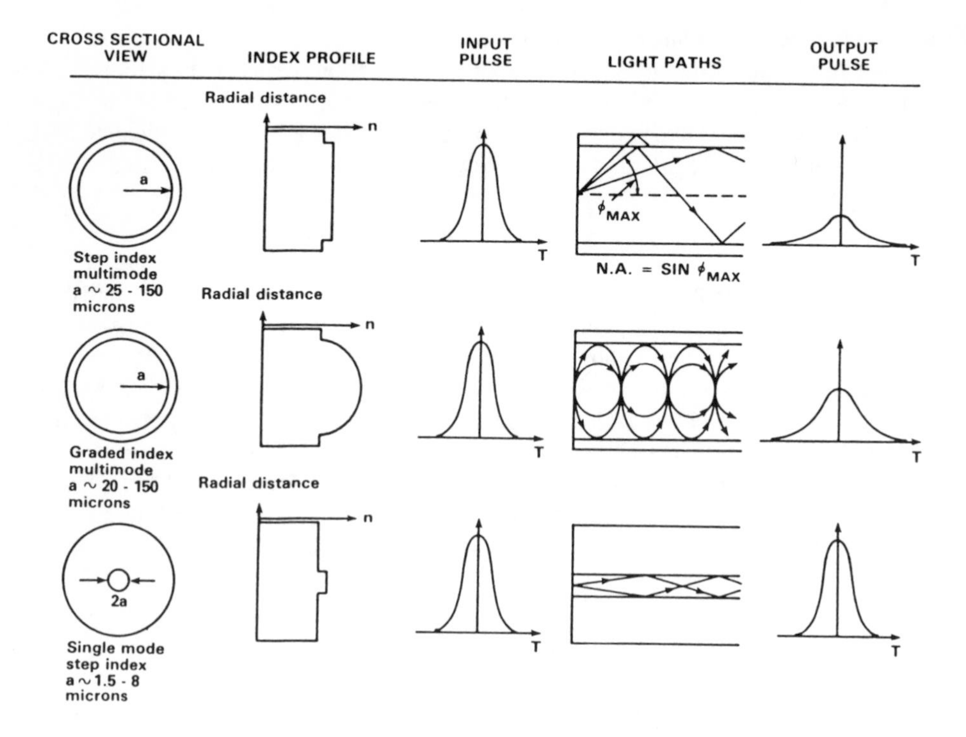

Figure 2. Optical fiber profiles and propagation patterns. (US Naval Research Laboratory, report under contract NOO173-79-C-0138, @ IEEE 1978).

travel a longer distance due to their reflection back and forth than low order modes which travel straight down the axis to the other end. Therefore we observe the phenomenon of "pulse spreading" or dispersion. This limits the amount of information we can send down a fiber because for very high data rates, pulses (as in a digital system) smear out into one another to an extent whereby they become indistinguishable. In the "graded-index" fiber this is compensated for to a certain extent. The index of refraction of the core is controlled in the manufacturing process to give it a gradient. The value of the index of the core varies from a maximum at the center of the fiber to that of the cladding at the edge (usually by the inverse square of the core radius).

When light in the high order modes travels in a graded index fiber, the light is refracted smoothly back to the core, rather than abruptly reflecting at the cladding. Since light travels slower in mediums of higher refractive index, the further from the axis the ray travels, the faster it moves. Therefore higher and lower modes arrive at nearly the same time at any given point along the axis, reducing dispersion greatly. In a single-mode fiber, like a step-index, the core is of uniform index, but the diameter of the core is so small that only the lowest order modes can propagate (the first TM_{00} and TE_{00}). The "single" mode has the least dispersion and therefore can carry great amounts of information without modal distortion.

Another major consideration for an optical fiber is absorption. The lower limit for the absorption of a glass or fluid material is the molecular or "Rayleigh" scattering limit. "[The] scattering of the light can be caused by the random arrangements of the molecules of the core material, which makes the material look slightly granular at light wavelengths. It is this [phenomenon] . . . that sets a lower bound to the loss of light in glassy or fluid materials." The attenuation decreases as the inverse fourth power of the wavelength.[6] Other absorption is caused principally by impurities in the glass. "If as little as one part per billion of certain metals (or even water) is included in the glass . . . the fiber will absorb a significant amount of light."[7] Some recent examples of this are shown in Figure 3. In this diagram, the absorption spectra for two fibers are shown with the Rayleigh scattering limit for each. This limit is a function not only of the material, but also varies as the Numerical Aperture (NA) of the fiber. The NA for a given fiber is the sine of the highest angle which can enter the end of the fiber and still be propagated down the wave-guide. Thus it is a measure of the coupling efficiency of the fiber to a light source.

Use of digital systems can aid optical fiber transmission by reducing the size of the signal-to-noise ratio (SNR) needed for a given signal's transmission on a specific system.

> For example, with a signal-to-noise ratio of 21dB only one pulse in a billion will be lost in the background noise [with a digital system]. For [a comparable] analogue

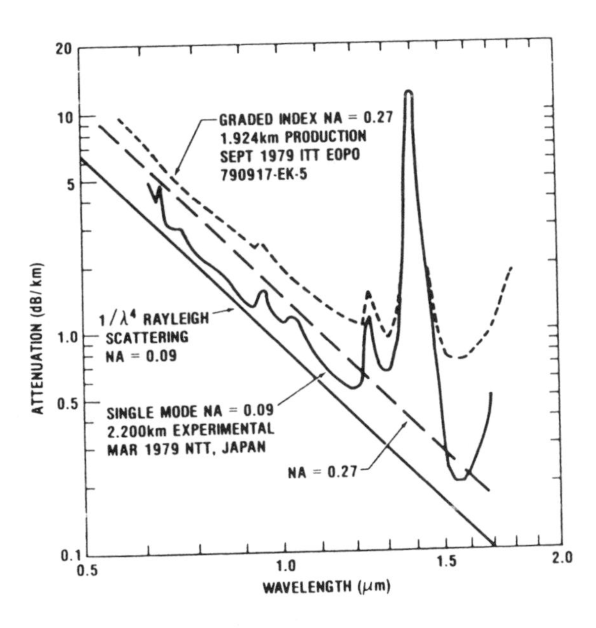

Figure 3. Attenuation of some optical fibers as a function of wavelength (λ). (Courtesy Robert J. Hoss, ITT Electro-Optic Products; GWU Continuing Engineering Education Course #541, October 1980.)

signal . . . any noise tends to distort the message; hence if the signal is to be satisfactorily reproduced . . . a signal-to-noise ratio of 60dB is needed.[8]

In the next section we shall cover some other points of interest from early theoretical work of Hondros and Debye in 1910[9] up through the development of light sources, detectors, cables, etc. in the recent past.[10]

The Historical Development of Optical Fiber Technology

There have been several innovations in the course of the growth of what has become fiber optics. These milestones have usually been preceded, sometimes by decades, by the theoretical background necessary to, or hinting at, the application of some newly discovered or developed technology to optical fiber communications.

As previously mentioned, the theoretical analysis performed by Hondros and Debye in 1910* was the ultimate basis for understanding the dielectric waveguide, but it was not until "the 1950's that optical fibres began to find practical application." These applications were generally as bundles which were used for illumination of instruments, field flatteners, coupling of plates for image intensifiers, photoelectric devices, copiers and nonsurgical diagnostic tools.[11]

The significant first step, however, came with the invention of the laser by Shawlow and Townes in 1958.** This led to the speculation mentioned in the introduction about using the wide bandwidth of the laser for communications. It was soon found that despite the directivity and power of the laser beam, attempts to use them in a manner akin to microwave relay links were frustrated by smog, snow, rain, fog and dust. This situation soon led to the realization that a protected environment was needed for transmission. This, in turn, led to the

> consideration of the use of conduits, possible evacuated pipes, for the sending of light beams from one place to another, thereby providing a controlled atmosphere. If necessary such conduits could be fitted with lens and mirror systems to provide path redirection.

Servomechanisms could be designed to compensate for environmental changes (and could be made to work) but would be so cumbersome as to be economical only for very high traffic routes.[12] A variation on this scheme was tried in the mid-1960s. This

---

*D. Hondros and P. Debye, "Electromagnetische Wellen und Dielektrischen Drahten," *Ann. Physik*, Vol. 32, (1910), pp. 465–470.

**A.L. Shawlow and C.H. Townes, "Infrared and Optical Lasers," *Phys. Rev., Vol. 12*, (Dec 1958), pp. 1940–1948.

> was to form what is called a gas lens. The scheme was to fill the conduit with gas, but to provide a radial temperature gradient by means of a suitable arrangement of heater elements. This radial temperature gradient in turn gave rise to a radial gas density gradient and hence a radial gradient of the refractive index. Such a radially graded refractive index could provide a waveguiding action for a light beam travelling more or less along the axis of the gas lens.[13]

While experiments, such as that by Christian, Gouban and Mink in 1967*, established the technical correctness of this approach, it was again likely to be an extremely cumbersome, unreliable application practice.

Also used in the late 1950s and early 1960s for short range transmission were ''light pipes'' in which the light is guided principally by internal reflection along internally silvered, hollow tubes or along plastic rods. ''Such schemes, while useful in many equipment applications, were not regarded as practical for long-distance communication because of the relatively high attenuations they exhibited.''[14] A major turning point occurred in the mid-1960s:

> In 1966, Kao and Hockam of Standard Telecommunication Laboratories in England, point out in a famous paper that the attenuation found in glasses employed for optical fibers was not a basic property of the material, but was produced by the presence of impurities, mainly metallic ions. Since the intrinsic material loss, essentially determined by Rayleigh type scattering . . . is very low, reduction of the impurity content would allow attainment of much lower losses than those typically encountered.[15]

Interest in the clumsier ''bulk'' systems of the 1950s and 1960s began to diminish when the use of glass fibers as a guiding medium began to gain notice. The older systems were economical only for the transmission of extremely large quantities of information. ''Size, weight, handling ease, flexibility and cost were clear advantages of such small dielectric waveguides . . .''[16] After Kao & Hockam's paper, investigation found that some of the best optical glasses of the time had attenuations in the thousands of dB/km. What was needed for a practical communications system were fibers with an attenuation of only a few tens of dB/km, but this implied an enormous improvement in glass technology with materials of unheard of purity, but Kao felt it might be possible. ''He was encouraged in his belief by measurements on pure silica that exhibited losses [in this range].''[17]

Kao and Hockam's paper sparked an intensive research program by the British Post Office for glass purification and the study of fiber transmission problems. Soon other organizations in the U.S., Japan and Germany followed suit with their own programs.[18]

While others were investigating the purity of materials, still others were investigating fiber structure.

---

*J.B. Christian, G. Gouban and J.W. Mink, ''Further Investigations with an Optical Beam Waveguide for Long Distance Transmission,'' *IEEE Trans. Micr. Theory and Tech., Vol. MTT-15, No. 4* (1967), pp. 216–219.

The monomode fibre [sic] was first considered as a suitable medium for transmitting large amounts of information. Multimode fibers were soon taken into consideration and, in 1968, an analysis of the refractive index grading necessary for minimizing model dispersion was carried out [by S. Kawakmi and J. Nishizawa in Japan].[19]

The key issue then, in the late sixties, was that the limit for economic feasibility lay in reducing the fiber attenuation to less than, or equal to, 20 dB/km. N. Lindgren, in his 1970 review in the Proceedings of the IEEE, assessed that existing fibers were not in a position to meet this goal.[20] He was proven wrong, however, in that same year.

The first really major breakthrough from an optical communications point of view was made by scientists at Corning Glass Works when they developed fibers based on high silicas. In 1970 they* announced fibers with attenuations as low as 20 dB/km; subsequently they were able to improve this figure to 4 and then 2 dB/km. [R.D. Maurer in 1974].[21]

Corning's method for fiber manufacture consisted of depositing a thin layer of very pure silica inside a fused silica tube. This ''mandril'' was then heated and collapsed to a solid rod which was drawn into a fiber. This forms ''the basis for most modern low-loss fibre-production technology.''[22]

At this time the analytic properties of the glasses people were working with were little known. This was so much the case that a major portion of the work of categorizing the refractive index as a function of temperature and composition had to be determined experimentally.[23] So little was known about the optical properties and the techniques to measure levels of impurities that ''one of the spin-offs from this glass program has been an increase in the sophistication of related analytical chemical techniques. Neutron activation analysis . . . has enabled the detection of levels down to a few parts per billion.''[24]

Corning's breakthrough encouraged others in other countries to enter the optical fiber development field. This stepped up concentration on fiber development led to the proposal, in 1974, by J. Stone to use quartz capillary tubes filled with a liquid, to serve as core, for use as a communications fiber. The losses, in practice, were on the order of 8 dB/km in near-infrared and the capillaries were useful for experimentation as the cores could be changed.[25] The announcement by R.D. Maurer [previously mentioned] in 1972 of fibers of high silica core material and losses of 4 dB/km in effect rendered liquid core fibers obsolete almost immediately.[26]

A substantially new achievement took place in 1976. Japanese researchers of NTT and Fujikara cables succeeded in fabricating a fiber, with a very low OH content and a minimum loss of 0.47 dB/km, very close to the intrinsic material loss. The only possible improvement at this stage was the exploitation of longer wavelengths, and, in fact, an attenuation as low as 0.2 dB/km has recently [1979] been reported for a

---

*F.P. Kapron, D.B. Keck, R.D. Maurer, *Appl. Phys. Lett.*, 17, (1970), p. 423.

multimode fiber at 1.55 μm, which seems to be the minimum attainable loss for doped silica fibers.[27]

At the same time that fibers and materials were being developed, work was going on with light sources and detectors. The successor of the fibers stimulated the search for components of appropriate size and power to allow their economic use in lightwave communication systems. Semiconductor light emitters and photodiodes appeared to be the most promising devices. Development progressed from the first cryogenically cooled laser diodes in 1962 and light emitting diodes (LED's) in 1963 through the first heterojunction lasers (HJL) in 1968 and continuous wave (CW) HJL's in 1970. By 1973 lasers with a life of more than 1,000 hours had been developed and extended to 7,000 hours by 1977. By 1979 devices had become available with expected lifetimes of 100,000 hours. In another breakthrough, in 1971, Buruss of Bell Labs had developed, in small area head emitting and high radiance, a small area LED (50 μm diameter) particularly suited to fiber optics.[28] Since 1979 lasers for continuous operation (CW) at room temperature with lives of 2,000 hours have been commercially available.[29] "As far as detectors are concerned, silicon PIN diodes and avalanche photodiodes with very good characteristics have been developed throughout these years without any particular difficulty . . ."[30] but a satisfactory low noise detector at longer (1.3μm) wavelengths is still in development. The other portions of fiber optic systems have been developed, fabricated, tested and field tested in many countries and are now available in commercial form.[31]

CONCLUSIONS

> The future of optical fiber communications looks very bright. There is no longer any serious obstacle to the introduction of such systems for the various applications which have been envisioned from the beginning . . . This is essentially due to the strikingly fast progress which has characterized this research field . . .[32]

and the attractiveness of optical fibers as a transmission medium. New materials are being studied to extend transmission distances by going to very long (3.5 − 5.5 μm) wavelength materials. It is projected that such materials as TlBr and $ZnCl_2$ may have attenuations in this region of 0.001 dB/km or hundreds of times less attenuation than "conventional" fibers.[33] This means that repeaters could have spacings of thousands rather than tens or hundreds of miles.

Fiber optic communications have become so attractive recently that within the last few months of 1983, AT&T and MCI each announced plans to install "massive new long-haul trunk systems using single-mode fibers. Both systems will operate at data rates above 400 Mbit/s—above the 270 Mbit/s T4 data rate which has been the highest level in the North American telecommunications hierarchy." AT&T plans to have installed 480,000 km of fiber over a 16,000 km route by 1995, the largest link of which shall handle 150,000 simultaneous voice channels.[34] It is estimated that the investment in hardware and installation shall cost AT&T and MCI $100 million each for their respective installation.[35]

To define the current state-of-the-art . . .

> Records for largest and highest speed repeaterless fiber transmission are established and surpassed recurringly. At present [September 1983], the record is the 420 Mbit/s transmission over 199 km, announced by Bell Telephone Laboratories . . . Two other groups, British Telephone Laboratories and Nippon Telegraph and Telephone, have also demonstrated repeaterless transmission over more than 100 km.[36]

Current records for attenuation are led by Corning Glass Works with a fiber attenuation of 0.16 dB/km at 1.55 μm wavelength and Bell Telephones' 119 km link with an attenuation of 0.27 dB/km including splicing and connector losses at a wavelength of 1.55 μm.[37]

New areas opening for application are undersea cables, military use, heavy traffic routes and long-haul land lines, and local area networks.[38]

For the point-to-point link category, fiber-optic land lines have some advantages over satellites. First, there are fewer satellite "slots" or parking orbits for geosynchronous purposes. Second, satellite radio-communication links are vulnerable to interception. The relay time for a satellite transmission is significant compared to an optical fiber line and, with 1983 technology, the fiber-optic line's "cost per circuit-month" is less than that of a satellite for distances less than 800 km (less than $300/circuit-month).[39]

Thus, we have surveyed the technology of fiber optics. Clear advances such as the discovery of the laser, the suggestions by Kao and Hockam, and the breakthroughs by Corning Glass and Nippon Telegraph being implemented and used as a basis for the next advance. Cooperation in this international endeavor has demonstrated how these innovations have sparked other researchers to channel their thoughts to new applications or techniques and in turn become innovative and institute the changes and developments they conceive to inspire and jog others' thoughts and ideas as to how something can be employed. On the other hand, we have also seen where this innovative, interactive process has broken down. On the whole then, this scientific, supportive, cooperative effort has resulted in the removal of a new, emerging and rapidly changing technology from the technical journal and laboratory to an increasing number of applications in today's world. We now do new things and old things better ways.

REFERENCES

1. W.S. Boyle, "Light-Wave Communications," *Scientific American, vol. 237, no. 2* (August 1977), p. 40.
2. Ibid.
3. J.S. Cook, "Communication by Optical Fiber," *Scientific American, vol. 229, no. 5* (November 1973), p. 28.
4. Ibid.

5. B. Costa et al., [CSELT], *Optical Fibre Communication* (New York: McGraw-Hill Book Company, 1980), p. 2.
6. Cook, "Communication by Optical Fiber," p. 29.
7. Ibid.
8. Boyle, "Light-Wave Communications," p. 44.
9. Costa, *Optical Fibre Communication*, p. 2.
10. Boyle, "Light-Wave Communications," p. 44f.
11. Costa, *Optical Fibre Communication*, p. 1.
12. A.G. Chynoweth, "The Fiber Lightguide," *Physics Today, vol. 29, no. 5* (May 1976), p. 28.
13. Ibid.
14. Ibid.
15. Costa, *Optical Fibre Communication*, p. 2.
16. Ibid.
17. Chynoweth, "The Fiber Lightguide," p. 28.
18. Costa, *Optical Fibre Communication*, p. 2.
19. Ibid., p. 2f.
20. Ibid.
21. Chynoweth, "The Fiber Lightguide," p. 30.
22. Costa, *Optical Fibre Communication*, p. 3.
23. Chynoweth, "The Fiber Lightguide," p. 29.
24. Ibid.
25. Costa, *Optical Fibre Communication*, p. 3.
26. Ibid.
27. Ibid., p. 4.
28. Ibid., p. 3.
29. Ibid., p. 4f.
30. Ibid., p. 5.
31. Ibid.
32. Ibid.
33. Ibid., p. 6.
34. D.A. Duke, "The Fiberoptic Industry in 1983: A Status Report," *Laser Focus, Including Electro-Optics Magazine, vol. 19, no. 9* (September 1983), p. 155.
35. Ibid., p. 156.
36. Ibid.
37. Ibid.
38. Ibid., p. 159.
39. Ibid., p. 162.

Application of Fiber-Optic Technology in Security Systems

H. Güttinger
G. Pfister

Abstract. The utilizations of fiber-optic technology in security systems are discussed. Of main interest are gas, intrusion and fire alarm systems. While many of the intrinsic features of fiber-optic technology are ideally suited for application in these systems, component and installation costs would have to drop further before security technology makes substantial use of fiber optics. In some cases like data transmission, however, it may already be justifiable to resort to fiber-optic technology. A review of the patent literature on fiber optics in security and the design of fiber-optic ionization and light-scattering detectors are described.

INTRODUCTION

Fiber optic technology is commonly associated with modern communication. Indeed, the astounding progress which has been made in the improvement of the transmission properties of the fiber, the perfection of the associated components, and the significant reduction of production costs has opened the door to large-scale commercialization of optical transmission technology. For communication, low attenuation, and a high bandwidth • distance product at lowest cost are crucial parameters, in all of which optical fibers outperform conventional transmission media such as waveguides or coaxial cables. In addition to these favourable properties, fiber optic technology offers a number of features which makes this technology attractive for application in areas well beyond that of pure high rate data transmission. This trend, of course, is promoted by the decreasing prices of fiber optic cables and components and the increasing

1983 International Carnahan Conference on Security Technology, Zurich, Switzerland, October 4–6, 1983.

availability of the technology. One area of interest concerns the introduction of fiber optics in security technology, which is the subject matter to be discussed in this paper.

Several features intrinsic to fiber optics can be put to use optimally in security systems.[1-4] These are

- absolute immunity against electromagnetic interference
- intrinsically safe in explosive environments
- no galvanic connection, i.e. no ground loops, perfect insulation
- fibers cannot be easily detected (in walls, etc.)
- fibers are not easily tapped
- no cross talk
- regulations for electrical wiring do not apply to installation of optical fibers
- large bandwidth • distance product
- low volume and weight

In order to illustrate the point, consider a large-scale fire and intrusion alarm system installed in an industrial building. Typically these systems can be very elaborate with many thousands of detectors mounted in locations ranging from pleasant office to very harsh factory environments. Equally complex is the routing of the associated power supply and data lines. The security system has to reliably detect hazardous conditions, for instance the onset of a fire or the presence of an intruder and signal such a condition appropriately. Since hazardous conditions should occur very infrequently, the security system should remain unnoticed for most of its lifetime. This brings us to the very important question of false alarms, i.e. the erroneous indication of a hazardous condition. False alarms are not only a nuisance to the people resident in the building but they also tend to promote negligence which could be very costly in case of a real hazard. False alarms can have a variety of causes among which malfunctioning of detectors, cross sensitivity of detectors and electromagnetic interference on detector and data lines are the most prevalent ones.

Some of the detectors may have to be installed in explosive environments. This is particularly true for detectors of explosive gases, for instance methane or hydrogen. In order to meet the regulations for safety, very elaborate electronics and shielding are often necessary which lead to substantial cost increases.

In many cases, also, it is desirable that the security system cannot be located and/or that data lines cannot be tapped. Again, with conventional electronic technology these conditions are met only with difficulty and additional cost.

In present day systems the detectors are installed in a two wire serial loop. So far the narrowband-width of such a system has not been a limiting factor because only binary data representing the status alarm yes/no of non-identified detectors were transmitted. However, the bandwidth of the two wire system will become increasingly limiting as one begins to change to systems in which individually addressable detectors send out analog signals representing the momentary detector status.

It is evident that many of the above mentioned shortcomings of present day security systems can be overcome by introducing fiber optics. Clearly, the most attractive system would be one which has no electronic components and no electrical wires distributed in the building. The entire system would be purely optical, i.e. sensor units and transmission lines. The signals transmitted via optical fibers would be analyzed in a remote central unit.

While data transmission, for instance between local control stations and a main computer or between detectors and local control stations, can be considered to be state-of-the art, the realization of fiber optic detectors is far from complete. In fact, fiber optic sensors for security applications are only now beginning to be investigated systematically. Furthermore, various components (connectors) are still too expensive for the installations required for distributed networks typical of security systems. Nevertheless, it is to be expected that for specific applications such as monitoring explosive environments, laboratories or high voltage power stations, fiber optic technology will gradually replace conventional systems.

It is the purpose of this paper to discuss present research and development of fiber optic sensors and systems for security applications. In Part 2 a very brief review of transmission in optical fibers will be given. Part 3 describes various concepts for implementing fiber optic technology in security systems. Part 4 contains a review of the patent activity related to the application of fiber optics in security together with some relevant developments of sensor research. In part 5 we present concepts for the realization of a fiber optic ionization and a fiber optic light scattering detector. The paper ends with a conclusion in part 6.

OPTICAL TRANSMISSION

Transmission of light along a fiber is based upon the principle of internal reflection. A beam of light launched into the core of a fiber is [confined to propagate] within the core if the angle of incidence at the boundary between the core and the surrounding cladding is smaller than the critical angle of total internal reflection $\theta_c = \sin^{-1}(n_{clad}/n_{core})$ where $n_{clad} < n_{core}$. Under these conditions the attenuation of light along the fiber is determined by the absorption and scattering of the light in the core and the cladding (which acts upon the evanescent field). High-quality silica fibers can be fabricated with an attenuation of only 0.3 db/km at the optimal optical wavelength of 1.3 μm.[5] While these low attenuations are crucial for long distance transmission, security systems generally do not require very long transmission lines. For most such applications cheaper fibers of a lesser quality will be sufficient. In particular, the much cheaper plastic optical fiber shows great promise for this application. Currently available plastic fibers have an attenuation of 300 db/km at the optical wavelength of 660 nm which allows for data transmission over distances up to 35 m without requiring repeaters.[6] There is intensive activity to reduce the loss into the 10 db/km range.[7]

The propagation of light waves in optical fibers is described in terms of mode analysis.[8,9] Accordingly, in a monomode fiber only the fundamental mode can be propagated. Because of the short wavelength of the propagation light (~ 1 μm) the core diameter of the monomode fiber is only a few microns. Monomode fibers are of first interest for long distance transmission of high data rates. The state-of-the art bandwidth for a 20 km link at 850 nm is 1.25 GHz and the cost of the fiber per 1

MHz • km is $0.05. By way of comparison a 50 GHz waveguide with a loss similar to that of the optical fiber (< 1 db/km) has a bandwidth of 100 MHz for a 20 km link at a cost of $ 100.- per 1 MHz • km.[10]

Monomode fibers can be used to construct fiber optic interferometers which form the basis of extremely sensitive fiber optic sensors, for instance for temperature, pressure or rotation (gyroscope). This is a very active area of research and development.[11,12]

Inspite of their outstanding performance record, monomode fibers are unlikely to be used in security systems within the next few years. They are relatively difficult to handle and the overall price, including components, is relatively high because of the small tolerances required. Most importantly, the wide bandwidth cannot be utilized in security systems which have comparatively low data rates, and the required sensitivity for the sensors is relatively low. Both these features can be satisfied with the more readily available and cheaper multimode fiber technology.

In a multimode fiber, hundreds of modes are transmitted simultaneously. In this case, the different propagation times of the various modes limit the bandwidth significantly but it remains sufficiently high for most applications. A typical telephone grade multimode fiber with an attenuation of < 3 db/km at 850 nm has a bandwidth of 50 MHz for a 20 km link at a cost of $2.30 per MHz • km.[10] If higher attenuations can be tolerated, the much cheaper plastic fibers may be used. At a bandwidth of a few kHz a feasable length for a 300 db/km plastic fiber is 150 m which is sufficient for data transmission in most security systems.

SYSTEMS AND SENSOR CONCEPT

In a conventional security system (for instance a fire alarm system) the building to be protected is divided into zones. For each zone a two to four wire serial loop with up to ~ 100 detectors is installed. The detectors are threshold devices with one or several alarm thresholds. The detectors are not identified, i.e. an alarm situation can be associated with a zone only.

There is a current trend to improve these conventional systems. Schemes are being proposed which allow identification of each detector and transmission of analog rather than threshold signals.

The introduction of fiber optic technology into security systems is likely to proceed in various stages. The obvious first application is the use of fibers for data links between local and central control panels. With current technology, the development of either multimode or monomode optical data links is straightforward. At a next stage one might consider the transmission of the analog signal of individual detectors via optical fibers. For smoke detectors, the bandwidth of the conventional two wire transmission technology limits the number of detectors to some 50 units if analog rather than binary data are sent. This number is even lower if the much more complex signals of intrusion or gas detectors have to be transmitted. Obviously, that shortcoming is resolved with fiber optic technology. Serial systems with conventional electrical detectors but with fiber optic transmission can be realized with commercially available combined electrical/optical cables. There will be some time, however, before cost-effective optical couplers, which have to be installed at each detector site, will be available and, furthermore, before technicians are trained to install these systems routinely and economically.

Ideally, of course, the entire security system would be based on fiber optic technology, i.e. the detector is realized in optical technology as well. Fiber optic sensors are, indeed, an area of active research and development.[11,12] There are essentially three types of fiber optic sensors which are being investigated:

(i) Using multimode fibers, the light is brought to the point of measurement where it is coupled out of the fiber, brought to interaction with a medium, which changes its optical properties as a function of the variable to be measured, and is then returned via the same or a second fiber. For these fiber optic sensors, the task essentially reduces to that of identifying and studying materials which change the optical properties appropriately. For example, in the optical microphone the light emerging from a first fiber is reflected off a pressure sensitive membrane and launched as pressure modulated beam into the second (returning) fiber.[13,14] A commercialized fiber optic temperature sensor is based upon the temperature-dependent fluorescent emission of doped rare earth crystals.[15] A fiber optic smoke detector has been proposed in which a first fiber brings light to the point of interaction with the smoke and a second fiber picks up the light scattered by the smoke aerosols.[4,16-18] Furthermore, radiation due to flames or sparks can be focussed onto the polished end of a fiber and detected by a remote control unit. Detectors based on this principle were suggested for the detection of fires or sparks in high-voltage environments like power stations, transformers etc.[19-21] There is an unlimited number of ways fiber optic sensors of this type can be constructed. This design is very attractive, also, because the sensor can be operated with conventional light sources or LEDs. For most applications in security, the sensitivity of the sensors achieved with this design is sufficient.

(ii) Alternatively, sensors can be constructed with the fiber itself being the sensing device. The parameters influencing the propagation properties of the light are mainly temperature and pressure. More recently, schemes were proposed in which the evanescent field is interacting with a cladding synthesized to be sensitive to gases.[22] By using monomode fibers, fiber optic interferometers of the Mach-Zehnder principle can be built, in which light from a laser source is launched into two fibers, a reference and a measurement arm.[11,12] The light is then combined and launched into the returning fiber. The resulting interference effect is a function of the T, P-, . . . -dependent propagation parameters of the measuring fiber. Because the length of the measurement arm can be chosen, the sensitivity of these devices can be made exceedingly high. A different type of fiber optic interferometer is the multimode fiber itself. The different modes superimpose as they propagate along the fiber resulting in a pattern of dark and bright spots, the speckle pattern, when the light emerging from the fiber is cast onto a plane as far field interference pattern. The speckle pattern is very sensitive to deformations which occur when pressure is applied along the fiber, for instance, by stepping onto a fiber buried underground.[23]

(iii) A combination of electronic and fiber optic sensor can be realized if the electrical power consumption of the sensor is low. For these detectors, light is guided via a fiber to the detector where it is converted into electrical power. The electrical output of the detector is converted back into an optical signal which is coupled into the returning fiber. Using this principle a telephone [24] and a ionization smoke detector[22,25] were powered.

Fiber sensors with non-fiber transducers (i), (iii) are best combined in a star rather than a serial system because the coupling loss between subsequent fibers along the loop is too large to make a serial installation attractive. Some types of line sensors (ii), on the other hand, can be utilized in a serial loop, for instance as a sequence of lengths of a multimode fiber which are pressure sensitive or which have the cladding removed to allow radiation to enter the core of the fiber.

REVIEW OF PATENT ACTIVITIES

The patent literature on data transmission in security systems is mainly concerned with the security of the data itself and the detection of interference on the line caused by sabotage or tapping.[26-28] Fibers are not easy to tap but by bending the bare fiber the light escaping from the core can be detected by a photomultiplier tube. This already quite difficult operation can be made even more difficult by various means described in the patent literature. For instance, the fiber can be coated with a highly stressed frangible glass which breaks upon bending the fiber.[26] Another method is to inject the information at a low and a masking signal at a high numerical aperture.[28] Microbending dominantly interferes with the higher modes in the fiber thus revealing the masking signal only. In addition, the wide bandwidth of a fiber optic transmission line allows very complex encoding schemes and multiple redundancies both of which are essential for secure data transmission.

Many proposals in the patent literature deal with fiber optic line sensors, i.e. lengths of fibers which are made sensitive to certain external stimuli.[23,29-52]

The pressure-sensitive speckle pattern emerging from the multimode fibers has been described in the preceding paragraph. This effect is proposed for the detection of intruders stepping onto a buried multimode fiber.[23,36,37] Other schemes make use of the fact that the breakage of a glassfiber is readily observable by a substantial decrease of the transmitted light. For instance, fibers incorporated into fences[31-35] or fibers glued onto window glass[41,42] or other structures[44-48] were proposed to detect forcefull breakage. Still other schemes make use of (i) the coupling of light between adjacent fibers which can be changed by pressure[40], (ii) the bending loss induced by heavier objects such as cars, driving onto a fiber clamped between two corrugated stiff structures[38] or (iii) the transmission delay in a fiber used to construct a relaxation oscillator which stops working when fiber breakage occurs.[43]

Line sensors for the detection of heat and smoke where also proposed. Fibers stripped off the protective cladding over a certain length of fiber can pick up surrounding radiation thus providing the base for a flame detector.[23,30] The transmission of a fiber with interruptions of air spaces decreases when smoke is developed. This design was proposed for building obscuration detectors. Fiber optic heat sensors make use of the temperature dependent loss of the optical absorption of the fiber cladding[49] or the temperature dependence of the bending loss.[50]

FIBER OPTIC IONIZATION AND LIGHT SCATTERING SMOKE DETECTORS

Ionization smoke detectors are very low power devices. MOSFET and CMOS based detectors running off a battery consume less than 10 μW electrical power in their quiescence state. It is, therefore, possible to power such a detector via an optical fiber, i.e. light is converted at the detector site into the electrical power necessary to drive the electronic circuits. Even more, part of the light at the detector site can be passed through a low power optical modulator and, via an optical fiber, be returned back to the local control unit. The modulator modulates the returning light in response to the detector output. Using these basic ideas we have built optically powered ionization detectors which return threshold values or analog representations of the detector output.[4,22,25,53,54]

For the threshold device a conventional high intensity LED light source at 820 nm and a plastic-clad silica fiber were used. At the detector site the light was cast onto a solar cell and the resulting 0.30 V were transformed to 12 V with an efficiency of 3% (representing the ratio of the electrical power available at 12 V to the optical power output at the fiber end). The overall efficiency, electrical power available at 12 V to electrical power input at LED including a 100 m long fiber was 10^{-4}.

The modulator was a LCD mounted between the solar cell and the input of the returning fiber. A hole was drilled into the center of the solar cell through which a small fraction of the incoming light was coupled out and focussed via the LCD into the returning fiber.

In a further development, the detector was modified to allow transmission of signals representing the analog status of the ionization chamber voltage.[4,54]

Now, the power consumption of the electronic circuit could not be supplied from a high intensity LED. For this reason we used a high intensity laser diode driven by 6 A, 50 ns pulses at 1 MHz (5% duty cycle). For efficient conversion of the optical signal into electrical power four serially connected solar cells proved to be the optimal choice. Again, a hole at the junction of all four solar cells admitted a fraction of the output light via the LCD switch into the returning fiber. The output of the laser is coupled into a 400 μm core, plastic-clad silica fiber by a cylindrical lens and converted to the required DC level by a DC/DC converter. The voltage of the ionization chamber, the sensor output voltage, is fed into a MOSFET source follower and conditioned by an operational amplifier. This voltage is then converted into a time interval by comparing it with a linearly ramping voltage. The timing information is used to modulate the LED switch. The light returned to the central station consists of pulses of fixed period separated by intervals proportional to the sensor voltage. To compensate for the temperature sensitivity only, the turn-off portion of the LCD switching characteristic was analyzed.

The *light scattering detector* offers the possibility to realize a fiber optic detector with no electronics located in the detector head, i.e. the driving and processing electronics are situated in a remote control unit.[4,17,22,55] Light is brought to the detector site and focused into a specified volume of space. When smoke aerosols enter this volume, a small fraction (10^{-6}) of the light is scattered in all directions where the angular dependence of the scattered light intensity is described by the Rayleigh formula for small ($\lambda >> r$) and the Mie formula for large ($\lambda \sim r$) particle radii, r[56].

A LED pulses light with 50% duty cycle into the input fiber. In one design, the fiber in the sensor head is coupled into a SELFOC-microlens which focuses the light into a beam of ~ 2 mm diameter and ~ 3° divergence. The power in the beam is ~ 0.1 mW. Only ~ 0.1 μW are spread into angles larger than the 3° divergence cone. A double aperture is used to block excess light which would be superimposed onto the light scattered from the smoke particles.

The return fiber is equipped with the same SELFOC-microlens which is arranged at a scattering angle of twelve degrees. More careful design of the optics would allow even lower scattering angles. Lower scattering angles would be an advantage because the scattered light intensity increases substantially at angles close to the forward direction. The light collected by the output collimator is led back to the control panel through the second fiber. There, a very sensitive receiver amplifies the incoming waveform. The output of the receiver is then fed into a lock-in amplifier which receives its reference signal from the LED pulser electronics. The output of the lock-in amplifier can be used as an analog signal or trigger an alarm circuit when a certain voltage level at the lock-in output has been exceeded.

A few details of the design deserve a closer look. The LED clearly has to be a small area high flux device in order to be able to couple as much power as possible into the 200 μm core plastic-clad silica fiber. The commercially available devices already contain a microlens for easy coupling to the fiber. Ray tracing in the optical system is quite straightforward and well described in SELFOC lenses data sheets[57]. Because there is no light coming back in the absence of an alarm, the receiver for the light is not shot noise limited and the electronic noise of the receiver determines the overall noise. The circuit chosen was a current to voltage converter. Its noise is mainly determined by the Johnson noise of the feedback resistor and the input noise current of the operational amplifier. Numerical calculations of the rms value of the noise agreed well with measurements and resulted in a NEP of 10^{-14} W/$\sqrt{Hz}$. The lock-in amplifier was set to a bandwidth of 0.2 Hz.

Actual tests of the device in the fire test facilities at Cerberus allowed the direct comparison of the fiber optic detector with a regular optical smoke detector. The result showed that the fiber optic detector yields a larger signal than conventional detectors for black smoke. This can be explained by the smaller scattering angle. Because the black particles are bigger, the gain in scattering intensity by Mie scattering partially compensates for their absorption.

CONCLUSIONS

There are several areas where the intrinsic properties of fiber optic technology could be used advantageously in security systems. Predominantly, these properties are the immunity to electromagnetic interference, the possibility of safe and reliable data transmission, and the fact that fibers are intrinsically safe in explosive environments. Despite these very attractive features, it will be some time before the optical fibers will be widely used in security systems, the main reason being that costs for components and installation are still high. The transmission of data, however, is state-of-the art and should be considered when electromagnetic interference is the cause of false alarms.

Much effort is needed towards the development of reliable fiber optic detectors. While the light scattering detector was easily realizable as a straight forward extension of conventional technology, the development of other detectors is less straight forward. Substantial effort is needed towards the development of fiber optic gas sensors, in particular for explosive gases where the advantage of fiber optic technology would come to its full bearing.

With the prices of components still falling, one can expect that research and development interests in fiber optic technology will continue to grow from this already respectable level. There is hope that some of these efforts will also benefit future needs of security technology.

REFERENCES

1. T.E. Larsen, ''Automation for Safety in Shipping and Offshore Petroleum Operations,'' ed. A.B. Aune and J. Vlietstra (North-Holland Publishing Company), 1980, pp. 199–205.
2. A. Brewster, *Security Surveyor,* November, 1981, pp. 29–36.
3. J.D. Montgomery and F.W. Dixon, *Proceedings of the Carnahan Conference on Crime Countermeasures*, University of Kentucky, Lexington, Kentucky, May 13–15, 1981, 121–4.
4. H. Güttinger and G. Pfister, *IEE Conference Publication Series*, 221, 1983, p. 62.
5. D.B. Keck, *IEEE Spectrum*, March 1983, p. 30.
6. R. Crofon, *Fiber Optics*, Du Pont.
7. T. Kaino, K. Jinguji, and S. Nara, *Appl. Phys. Letters* 42, 1983, p. 567.
8. D. Marcuse, *Theory of Dielectric Optical Waveguides* (Orlando, FL: Academic Press, 1974).
9. Technical Staff of CSELT, *1980 Optical Fibre Communication* (New York: McGraw-Hill Book Company, 1980).
10. G. Lutes, *Laser Focus* 18(9), 1982, p. 115.
11. T.G. Giallorenzi et. al., *IEEE Trans. on Microw. Th. and Tech.*, MTT-30, 1982, pp. 472–511.
12. *IEE Conference Publication Series*, 1983, p. 221.
13. I. Fromm, and H. Unterberger, *Elektronik*, Nr. 26, 1979, pp. 52–4.
14. I. Fromm, *Frequenz*, 32, 1978, pp. 356–63.
15. K.A. Wickersheim, US 4 215, 1980, p. 275.
16. W. Schnell, J. Muggli and G. Pfister, US 4 288 790, 1981.
17. M. Pistor, H. Rohrbacher, and A. Schaffernak, DE 30 31 674 A1, 1982.
18. A. Miyabe, DE 3044 944 A1, 1981.
19. M. Russwurm, DE 30 17 144 A1, 1981.
20. C.E. Lindgren, DE 3031 517 A1, 1981.
21. J. Dorosz and R.S. Romaniuk, *Proceedings of the Int. Fiber Optics and Commun. Expos.*, 8, 1982, p. 176.

22. J. Muggli and G. Pfister, BE 881 812, 1980.
23. C.D. Butter, GB 2 046 437A; DE 30 11 052 A1, 1980.
24. R.C. Miller and R.B. Lawry, *Bell Sys. Sys. Tech. J.*, 58, 1979, pp. 1735–41.
25. H. Güttinger and G. Pfister, *Proceedings of the Second European Fiber Optics and Communications Exposition*, Cologne, 1981, pp. 55–6.
26. J.A. Rupp, US 4 174 149, 1979.
27. J.L. Goldberg and A.A. Sadler, EP 0 006 364, 1980.
28. P.D. Steensma, US 4 211 468, 1980.
29. D.W. Burt, GB 1 583 700, 1981.
30. D.W. Burt, GB 1 580 272, 1980.
31. J.A. Wyman, GB 2 038 060 A, 1980.
32. J. Ciordinik and A. Penzo, DE 30 29 712 A1, 1981.
33. R.V. Fletcher and R.E. Oseland, GB 2 062 321 A, 1981.
34. R.F. Bridge, US 4 307 386, 1981.
35. C.A. Kitchen, EP 0 049 979 A2, 1982.
36. G.D. Pitt and R.J. Williamson, GB 2 057 120 A, 1981.
37. S.J. Stecher, US 4 321 463, 1982.
38. J. Robieux, CH 627 573, 1982.
39. M.M. Ramsey, GB 1 497 995, 1978.
40. G.M. Forbes and R.J. Seaney, GB 2 091 874 A, 1982.
41. E.G. Eatwell, GB 1 602 112, 1981.
42. P. Bourely, FR 2 418 504, 1979.
43. B. Perren, DE 30 21 705 A1, 1981.
44. P.H. Edward and N. Parkinson, GB 2 077 471 A, 1981.
45. B. Allias, FR 2 452 559, 1980.
46. *New Scientist*, 24 September 1981, p. 798.
47. H.L. Ditscheid, EP 0 025 815 A1, 1981.
48. L.K. Thompson and W.H. Corbett, EP 0 033 867 A1, 1981.
49. K. Blotekjaer and G.T. Sincerbox, *IBM Tech. Disclosure Bull.*, 20, 1929, 1977.
50. M. Gottlieb, G.B. Brandt, and J. Butler, *ISA Transactions 19*(4), 55, 1980.
51. T. Chijuma and Y. Morita, US 3 805 066, 1974.
52. S.C. Peek, BE 881 994, 1980.
53. J. Muggli and G. Pfister, US 4 379 290, 1983.
54. W. Meier and H. Güttinger, *Helvetica Physica Acta* 1983 (Forthcoming).
55. J. Muggli and G. Pfister, DE 3 037 636 A1, 1980.
56. Van de Hulst, *Light Scattering by Small Particles* (New York: Wiley, 1957).
57. SELFOC lenses data sheets, Nippon Sheet Glass Company Ltd., Osaka, Japan.

PART VI

Security System Control

As stated earlier, technical security systems are composed of at least a sensor, an alarm or annunciator, communication link and control unit. Intrusion detection sensors and CCTV have been discussed in Parts III and IV. Part V presented information on the state of the art communication link in fiber-optic medium. This section, Part VI, completes discussion of intrusion detection systems by explicitly presenting a complex, but simple to operate control center and implicitly presents annouciation, which is an integrated function of the Sentrax II Control Console.

Control units can be very simple if systems contain only a couple of sensors. The control unit merely provides power, activates the system, resets it after an alarm, and integrates the sensors. The control function becomes extremely complicated in a large and complex facility, such as a nuclear power plant. Many and varied intrusion detection sensors, smoke and fire sensors, and other hazard sensors, as well as CCTV, must be integrated and communicated to a security operator so he or she can pinpoint the nature, extent and location of the danger. In addition, the operator must make response decisions accurately and rapidly to neutralize the danger. Then, the operator must communicate these decisions to those who need to respond to the danger.

These are among the many complicated functions performed by the commercially available control console, Sentrax II. It is currently in service on many diverse sites. The Sentrax II computers simplify the control functions by rapidly processing vast amounts of data and presenting appropriate response options to the security operator. Not only is the unit designed to facilitate decision making, but it is designed to be convenient and comfortable for security personnel to use.

The Sentrax II is also designed for severe environments to minimize equipment failure. The automatic self-diagnostic component precisely describes the malfunction to the operator. In the rare case when a malfunction occurs, it can be instantaneously repaired because of its modular construction.

SENTRAX II
An Integrated Security Display and Control System

Edwin H. Morton

Abstract. SENTRAX II is a display and control system which allows one operator to monitor and control complex internal and external security subsystems from a control console. The System makes full use of distributed microprocessor technology and interactive color graphic display to provide a 'user friendly' display and control capability. Both perimeter and internal security subsystems may be integrated by SENTRAX II.

Surveillance, assessment and communication activities are all performed from a custom-configured console which is designed for maximum operator efficiency. The system provides a full data logging capability. Audio and video recorders can also be integrated.

Key functional elements are duplicated to provide a high degree of availability and an uninterruptible power supply can be provided. The high-quality, color graphic display and control unit includes three or five buttons which are used to operate complex sensor/site configurations.

Sensor and control connections are achieved through relay closures or RS232-type serial interfaces. If required, the SENTRAX perimeter intrusion detection system may be incorporated directly into the integrated system. SENTRAX II is ideally suited for applications such as large commercial and industrial facilities, correctional facilities, power generating plants and defence facilities.

INTRODUCTION

This paper introduces SENTRAX II, an Integrated Security Display and Control System that has been designed and developed to meet a range of security requirements. SENTRAX II systems have been installed and operated at several locations in conjunction

1984 Carnahan Conference on Security Technology, University of Kentucky, Lexington, Kentucky, May 16–18, 1984.

with the SENTRAX perimeter Motion Detection System. Distributed microprocessor technology is used to control the wide range of inputs and outputs handled by the system. One operator monitors and controls complex internal and external security systems from a central console. Interactive, menu-driven color graphic displays present sensor information to operators clearly and simply. Complex "housekeeping" decisions are handled by the system and the operator is presented with only the relevant choices for all situations. Five push-buttons provide control of the critical functions of external surveillance and assessment devices.

PRODUCT DESCRIPTION

SENTRAX II is a powerful microprocessor-based Display and Control System which is designed to interface with a wide range of perimeter and internal security systems. It has been designed as a reliable, rugged and cost-effective method of integrating the security systems found in large industrial complexes, oil refineries, penal institutions, etc. Figure 1 shows a typical SENTRAX II configuration, illustrating its interconnection with various external sensors and assessment devices.

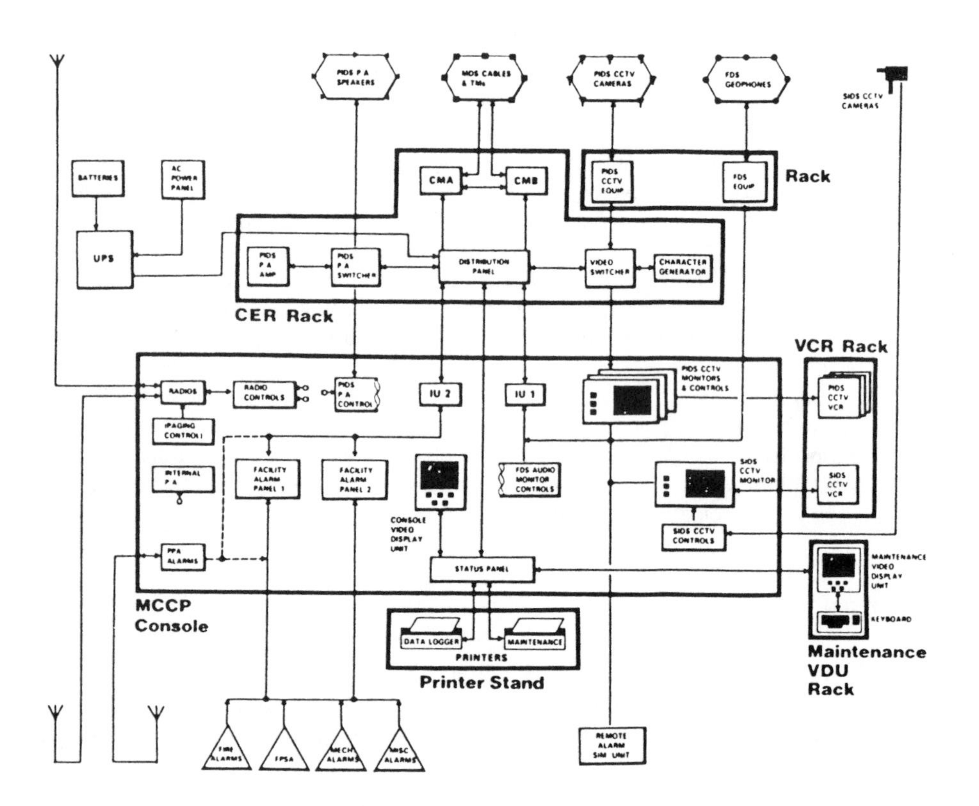

Figure 1. Typical SENTRAX II configuration.

An installation consists of the following components:

- an operator's console;
- a maintenance/backup display and control unit;
- a control equipment rack;
- a selection of recording and data logging equipment;
- an uninterruptible power supply.

The operator's console houses all of the controls needed to operate the surveillance and assessment systems integrated by SENTRAX II. A typical console contains the following items:

- CCTV monitors;
- a radio communications panel;
- perimeter/internal public address (PA) controls;
- SENTRAX II display and control unit;
- SENTRAX II backup system controls;
- internal security alarm panel(s);
- paging system controls.

In addition, the operator's console contains the interface equipment which connects the microcomputer and SENTRAX II hardware to conventional relay-closure/switch-closure inputs.

The *maintenance/backup display and control unit* is housed in its own rack, adjacent to the operator's console. This unit is identical to the display and control unit in the console and is activated from a console control in the event of failure of the main unit.

The *control equipment rack* may be located at a distance from the main console in a secure equipment room. The rack houses the main control computers, video switching equipment, audio switching equipment, PA amplifiers and main distribution panel.

Video cassette recorders and printers are housed in auxiliary racks or stands depending upon the layout of the control room.

An *uninterruptible power supply and battery unit* may also be located at a distance from the main console.

CONTROL ARCHITECTURE

Overall control of the system is resident in a pair of microprocessors, called Control Modules (CMs). These computers interface with the other elements of the system via serial, RS232-compatible links, or by one or more parallel collection of distribution points. A block schematic of the control architecture is shown in Figure 2.

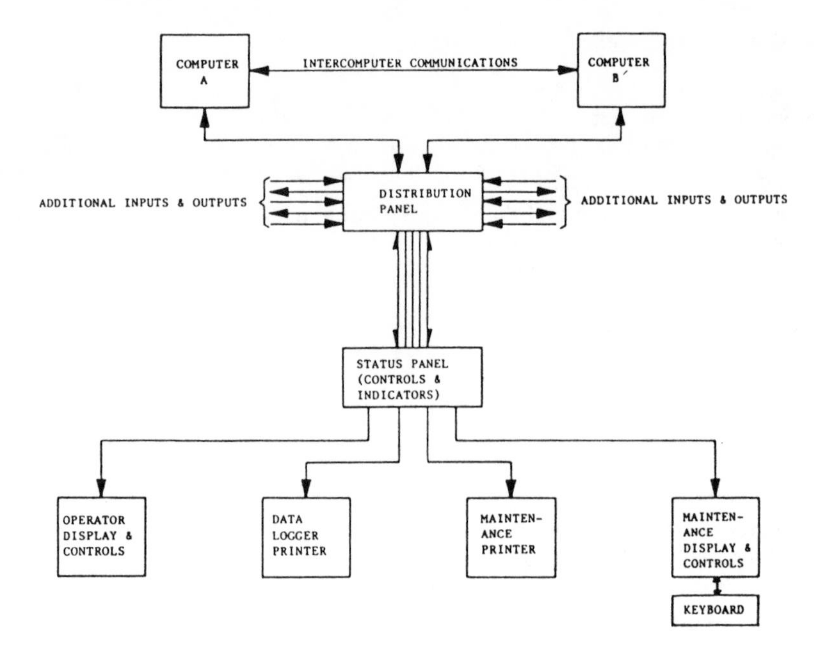

Figure 2. SENTRAX II control architecture block schematic.

In addition to two CMs, a pair of printers and a pair of display and control units are provided. Switches on the operator's console designate the primary computer, printer, and display and control unit. This configuration provides an extremely high level of reliability and availability.

The CMs are equipped with eight serial interfaces and several parallel input and output ports. The serial ports handle the following signals:

- communication between the CMs;
- communication with a video switcher;
- communication with up to three interface units;
- communication with an external data i/o port;
- communication with a time and date generator.

Each CM also has a direct communication port which provides a connection to a display and control unit and a printer. This configuration is shown in Figure 2.

If the SENTRAX II system is installed at a location equipped with a SENTRAX perimeter Motion Detection System, the coaxial cables interface directly to the CMs. Additional perimeter and internal alarm systems are connected through the interface units shown in Figure 1.

DATA ACQUISITION

The SENTRAX II System gathers data from the various internal and external alarm systems with an Interface Unit (IU). Each IU can be equipped with up to 256 collector points which respond to a logical change of state. SENTRAX II can handle up to three Interface Units. If required, an IU can be equipped with distribution points, in blocks of 16, to drive external systems. Each block of 16 distributor points replaces a block of 16 collector points, up to a maximum of 256 points.

SENTRAX II is programmed to respond to the collector inputs and to respond in a predetermined manner to close collector outputs, change the state of the Display and Control Unit, etc.

DISPLAY AND CONTROL

The display and control unit provides a simple and effective means for an operator to control a complex security system. Figure 3 shows the display and controls.

The display and control unit is a color graphics video display unit (VDU) which shows a scaled map of the perimeter. The map includes an outline of the buildings, the perimeter divided into sectors and CCTV cameras. When there is an alarm, the alarm sector is indicated by a red symbol. When a sector is masked (deactivated), it is indicated by a yellow symbol. At the bottom of the screen there is a text line which provides prompts and mode descriptions for the operator.

There are two VDUs: a primary VDU in the operator's console which is used for

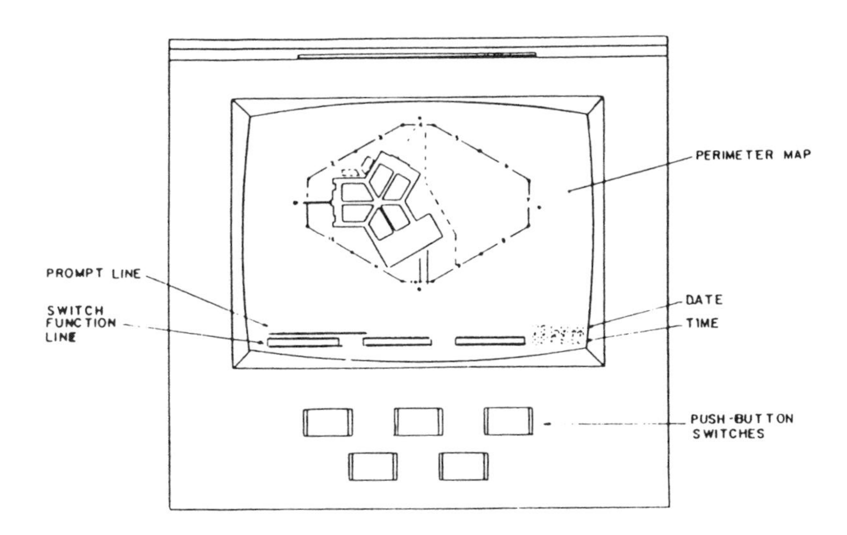

Figure 3. Display and control unit.

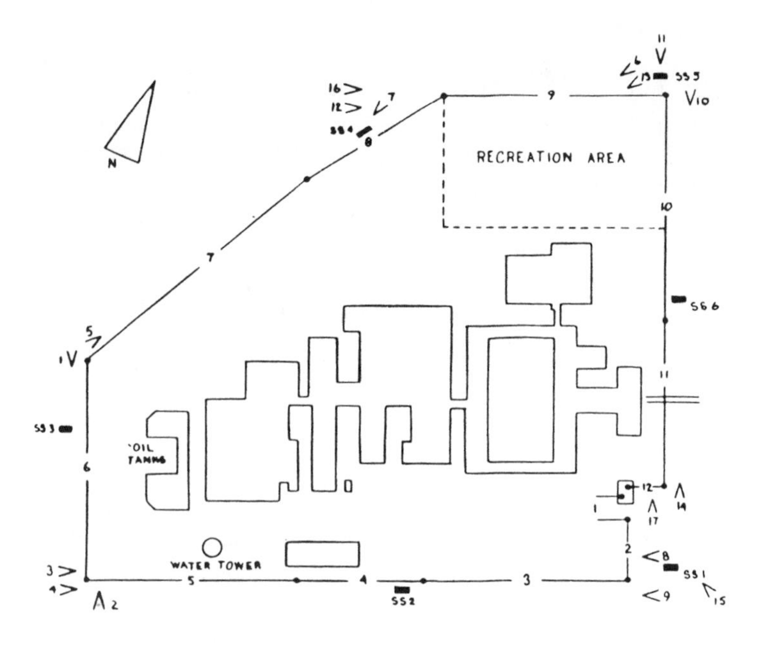

Figure 4. Map of perimeter.

normal operation; and a maintenance/backup VDU which is used for occasional maintenance and as a backup should the console VDU fail.

Display

The VDU is a graphic display terminal which shows the status of the perimeter in the form of a map, as illustrated in Figure 4.

Map Area

When no alarms are present, the map displays the following symbols:

- Perimeter—Light Blue (Secure), Yellow (Masked)
- Building Outlines—Dark Blue
- Additional Fences—Dark Blue (dotted)
- Towers—Dark Blue
- CCTV Cameras—White

Cursor

The operator instructs the system to process alarms, one sector at a time. The current sector is defined by a "flashing" sector number, called the CURSOR. The following equipment is "tracked" by CURSOR movement: CCTV Cameras, Perimeter PA System, and Perimeter Instrusion Detection Systems.

Prompt Line

The PROMPT LINE serves as an "operator prompt" and displays single-line messages indicating the mode in which the VDU is currently operating, the status of the system and abbreviated instructions. The PROMPT LINE appears in steady-state green or red, depending on the system status.

Switch Function Line

The SWITCH FUNCTION LINE indicates the function of the top row of buttons on the VDU. Each box corresponds to a button immediately below it. The text in the box defines the function currently available from that key. If there is no box over a button, the button will not be illuminated and no action will occur if it is pressed.

"Soft-Key" Push Buttons

The top three buttons are soft keys. Functions are defined by the SWITCH FUNCTION LINE. They are only active when illuminated.

Cursor Buttons

The lower pair of buttons control the movement of the VDU CURSOR. They are only active when illuminated.

Date Line

The DATE LINE displays the current date in green.

Time Line

The TIME LINE displays the current time in green. Following system power up, the TIME LINE flashes until the time has been reentered at the keyboard.

RELIABILITY AND MAINTAINABILITY

SENTRAX II is a commercial product designed for operation in severe technical and operational environments. The duplexed architecture of the system provides an extended Mean Time Between System Failure (MTBSF) of thirty-six years for the primary display control and processing elements.

Modular equipment, and the use of connectors wherever possible, reduces the time required to replace a module or card to a maximum of ten minutes.

Maintainability of the system is further enhanced by an automatic failure diagnostic capability. All SENTRAX II equipment is continuously monitored and failures are flagged for the operator at the display and control unit. The operator is informed of the type of failure and any corrective action to be taken. The diagnostic message may then be released from the display and is recorded on the maintenance printer. Messages can be displayed on the screen of the maintenance video display unit (VDU) for action by maintenance staff.

APPLICATION

SENTRAX II can be configured to meet a wide range of security and control applications. The display and control map and text may be configured to accommodate virtually any exterior or interior configuration. Various types of input/output protocols (RS-232, RS-422, opto-isolated, form C, etc.) can be handled through the selection of the appropriate interface cards.

Operator controls are mounted in a compact console and are grouped together to ensure ease of use. Following system design and definition, console configurations are adapted to meet individual requirements.

SENTRAX II is used for the display and control of perimeter and internal sensors in the following typical applications:

- industrial complexes;
- power generator stations;
- petro-chemical sites;
- penal institutions;
- strategic residences;
- air fields.

SPECIFICATIONS

To maximize the integration and control capabilities of SENTRAX II, a detailed specification is required. This would normally include the operation and electrical specifications of all systems to be integrated into, or controlled by, SENTRAX II.

A complete written definition of the operating interfaces and the anticipated responses of the system is required. A system specification is then prepared by Senstar which serves as a design yard stick for each installation. Detailed documentation for operators and maintenance technicians is supplied as required.

The following critical items should be defined when specifying a SENTRAX II system:

- external sensors: type, number, interfaces;
- internal sensors: type, number, interfaces;
- surveillance equipment: type, interfaces;
- assessment requirements: type, interfaces;
- communication requirements;
- anticipated response times;

- security levels of transmission media;
- maximum number of graphic maps required;
- quantities of text to be displayed;
- levels of system interaction.

INSTALLATION

Following an extensive ''in-plant'' ''burin'' and evaluation, SENTRAX II can be installed as a ''turnkey'' system to minimize disruption of day-to-day operation. Alternatively, if on-site security requirements prohibit the use of outside contractors, SENTRAX II may be installed easily by the end user because of its modular construction and use of connectors. In addition, map and text information may be loaded by either the end user or by Senstar personnel.

CONCLUSIONS

SENTRAX II systems have been installed at several locations across Canada where they are controlling and integrating the following systems:

- Perimeter Sensors (MDS and FDS);
- Perimeter CCTV;
- PA;
- Local and Regional Radio Communication;
- Perimeter Audio Monitoring;
- Data Logging;
- Internal Alarms: fire, mechanical, fixed point, portable, other miscellaneous;
- Video Cassette Recorders;
- Uninterruptible Power Supply.

The installations have been well received by the users. Careful attention to the human factor in the design and engineering of the system has minimized operator training requirements. Senstar is currently planning the installation of similar systems at an additional five locations within the next six months.

PART VII

Access Control

Access control or the differentiation between authorized and unauthorized persons is vital to most security systems. Someone or some process must differentiate these people by some means. There are essentially three means of doing this: by something the authorized person has, knows, or is. The first two are vulnerable; an imposter can steal a unique possession or gain knowledge. An imposter can also disguise himself or herself as the authorized person.

Certain attributes, however, are difficult to alter when imitating authorized personnel. These attributes can be identified by a variety of biometric measurement devices. Computer processing can rapidly compare the attribute configuration of those asserting to be an authorized person with the attribute configuration of the authorized person stored in the memory of the computer. There are biometric identity verification systems measuring a number of different attributes. There are a number of devices on the market which access figure geometry, palm geometry, voice patterns, eye retina patterns, the process of writing one's signature, or finger prints. Costs begin at $6,000.

Fingerprint assessment is used in the biometric identity verification system reported in the article of this section. The article describes the device and the application of the PIV-100 unit produced by De La Rue Printrak, Inc. in detail.

A Real-Time Positive Identity Verification System

Richard D. Capello
R. E. Hilderbran

Abstract. An innovative new real-time system has been developed for positive identity verification, using data extracted from the attributes of the human fingerprint as a unique personal identifier. The system provides a highly reliable answer to the question: "Is this person really who he or she claims to be?" It is based on the DEC Micro-11/73, running RSX-11M, with special hardware and software for processing the fingerprint images

INTRODUCTION

The rapid growth over the past twenty years of fraud, theft and terrorist activity has stimulated concern over protecting valuable resources from abuse. Recent publicity of "computer break-ins" to government data-bases by teenage "hackers" has raised general awareness of the dangers of unauthorized access to data.[1] In most applications, security boils down to the problem of verifying that only authorized individuals, or members of an authorized group, are allowed access to a protected item, location or service. In practical terms, this means verifying the identity and authorization rights of each applicant before granting access.

Potential applications for identity verification fall into four major categories.

1. Physical access control to computer rooms, bank vaults, secure buildings, nuclear facilities, military sites, etc.
2. Data access control, such as log-on to multi-user systems or networks, confidential database access, and subscriber data services.

1985 Carnahan Conference on Security Technology, University of Kentucky, Lexington, Kentucky, May 15–17, 1985.

3. Credit access control, such as retail point of sale, electronic funds transfer, automatic remote tellers, etc.
4. General identity verification for personnel time management, hospital operations, law enforcement, and international passports.

Today's market for identity verification is highly fragmented, with numerous competitors and many competing technologies. The most common methods are based on visual recognition (typically where personnel carry an ID badge with embedded photograph), on keys or magnetic cards of various types, and on memorized four-digit personal ID numbers. Today's technologies generally do not recognize a unique human trait or attribute, and can be compromised through alteration, duplication or theft of the badges or keys, or through human error. Warfel's book[2] provides an excellent survey of these issues. FIPS publication 83[3] gives a good outline of authentication techniques for access to computer networks.

POSITIVE IDENTITY VERIFICATION

Positive identity verification is focused on the issue of ensuring that "a person really is who he or she claims to be." Note how it differs from identification, where the person is unknown to the system, leading to a search of the database of the entire user population in an effort to match the person's attributes. Identity verification on the other hand, is a one-to-one comparison of attributes. The person first enters an ID number so that the system can retrieve one set of enrollment attributes from the database, and then match these with the "live" attributes presented. This fact clearly reduces the processing necessary, and makes feasible the design of a system that can give a "real-time" response at reasonable cost.

There are three basic methods by which the identity of an individual may be established[4]:

1. Something a person HAS, such as a key, badge or card
2. Something a person KNOWS, such as a password, PIN number, pass algorithm, or personalized facts
3. Something a person IS, a physiological attribute such as a finger or palm print, hand geometry, signature dynamics, voice print, or retinal pattern

It is readily apparent that the first two methods are not very secure by themselves. Keys or cards can be stolen, and passwords or PIN numbers can be learned through coercion or seeing them written down. Only personal physiological attributes provide reliable security, because of the difficulty of forging or transferring them to another person. The ultimate security is achieved by using personal attributes in conjunction with one of the first two methods.

Of all physiological attributes, fingerprints are the best suited to positive identity verification. They have the advantages of being unique for each finger of each person in the world, they do not change throughout the lifetime of the person, they are abundant (usually ten per person to choose from), they are easily presented to a machine, and no learning or practise is required.

FINGERPRINT TECHNOLOGY

Fingerprints arise from the formation of papillary ridges in the epidermal, or outer skin layer, of the fingers. These ridges serve the biological function of "friction ridges" in assisting our sense of touch, for the fingers are endowed with an abundance of nerve-endings.

Although the diversity of fingerprint ridge patterns had been known for thousands of years, having been used by the early Egyptians and Chinese, it was only toward the end of the nineteenth century that fingerprints came to be used systematically as a means of criminal identification. In 1894 the British Government adopted a method of classification and filing devised by Sir Francis Galton, later refined to become the "Henry System" employed by police forces throughout the world.[5]

The science of fingerprints as a means of personal identification is founded upon two basic precepts: their persistency, meaning that the pattern on each finger is permanent and unchanging (apart from growth) from before birth until decomposition of the skin after death, and their uniqueness, meaning that the ridge detail is never duplicated, i.e., no two people have ever been found to have identical prints.

For classification purposes, fingerprint pattern types are divided into the three general classes of arch, loop and whorl. These are further resolved into one hundred and fourteen sub-classes for more precise classification in the system used by the FBI.[6] Other systems based on syntactic pattern analysis have been proposed that result in fewer distinct classes[7,8], though these are primarily of academic interest. Irrespective of its classification, the unique identification of a fingerprint depends on the distribution of minutia points within the print. A minutia is defined as either an endpoint or bifurcation of a ridge, and is characterized by its position (X,Y) and angle within the print.

Fingerprint images can be captured by either indirect or direct means. The traditional, indirect method is to coat the finger with black ink and then roll it onto a white card, giving an image that is black along the ridges and white in the valleys, with some intermediate shades of grey. More recently, chemically coated cards have been used for "inkless" prints. As the skin of the fingers is rich in sweat glands, contact with any non-porous surface will leave a "latent" print that may last for several days. These "scene of the crime" prints are of immense importance in criminal investigation. The latent images are dusted with powder, or developed in a variety of ways, and then photographed.

For real-time analysis of fingerprints, a direct method of capturing the image is required. The well-known principle of frustrated total internal reflection (FTIR) is commonly used,[9] wherein an image is formed by the light that is not reflected from the underside of the surface of a glass prism on which the finger is pressed. New types of direct-contact sensor are presently under development, employing piezo-electric or pyro-electric materials, such as polyvinylidene fluoride, with CCD cell arrays,[10] or micro-mechanical structures of silicon.[11]

The automated processing of fingerprints involves extensive image processing. Figure 1 shows the four stages of an image as the requisite data is extracted. First a grey-level image is captured (1a) typically at a spatial resolution of 512 by 512 pixels by eight bits per pixel. This image is then equalized to give a uniform density range over the whole area, thresholded to a binary image of one bit per pixel, and enhanced to smooth out the ridge structure (1b). The ridges are then thinned to a minimal connectivity network (1c) from which the minutiae locations (1d) are extracted. The final minutia list, containing about 100 minutiae stored as three bytes each, represents a data reduction factor of nearly 1000.1 over the original grey-level image.

Fingerprints are matched against one another by a kind of correlation process

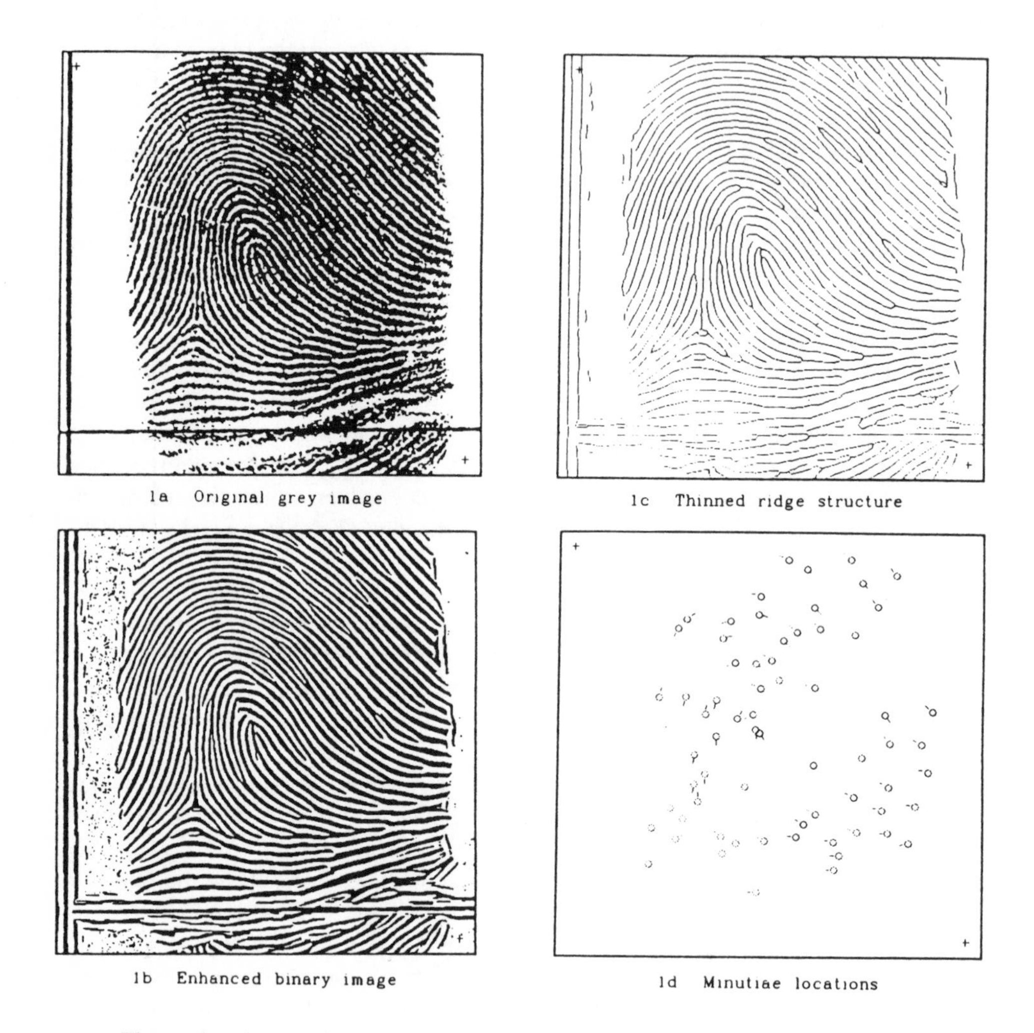

Figure 1. Stages of enhancement and data extraction from a fingerprint image.

between their respective sets of minutiae.[12] Within a specified local neighborhood of each minutia of the first print, a score is calculated from the number of minutiae of the second print having a similar angle. The total match score is then calculated as the sum of all these neighborhood scores. The algorithm tries many relative translations and rotations to obtain the "best fit" between the two prints. Figure 2 shows a plot of the matching minutiae of two similar prints.

SYSTEM ARCHITECTURE

De La Rue Printrak has developed a new product, known as the PIV-100, that mechanizes the process of positive identity verification, using an applicant's fingerprint as the unique personal attribute. This product has arisen out of work done some years ago at Rockwell International,[13] and has been expedited by Printrak's expertise and pre-eminence in the field of criminal fingerprint identification.

The PIV-100 is designed for the application of physical access control, i.e., verifying the identity and authorization rights of persons entering a protected room or facility. It will allow users, who have been enrolled into the system database, into the areas for which they are authorized. It will exclude, with a high degree of reliability, users from areas where they are not authorized. The PIV-100 has the secondary function of detecting and reacting to a variety of alarm conditions, ranging from sensors for fire, intruders or tampering, to such circumstances as a fraudulent access attempt or a protected area being filled to capacity.

Each point of entry into a protected area is termed a "portrait." At its simplest, this may consist only of a door or turnstile with an electric-release lock. In a very high security environment, such as a nuclear site, a portal may consist of an armored chamber with two interlocking doors. It may contain a wide variety of detection devices (for explosives, weapons, chemicals, nuclear materials, etc.) and communication with a guard station via CCTV and intercom. Some very sophisticated analysis of portal requirements for entry control systems has been carried out by Sandia Laboratories[14] for the U.S. Department of Energy.

Figure 3 shows the components of the PIV-100 system. Up to sixteen identification terminals may be connected to the system controller. The terminals may be of either a "table mount" or "wall mount" construction. The central processor is based on the DEC Micro-11/73. The system is controlled from an operator's station consisting of a control/display terminal, a printer, and a monochrome video monitor.

Each portal into the protected area is controlled by one PIV-100 terminal, which performs the functions of both identification, via a keypad into which the user keys his PIN number, and verification via a finger scanner onto which the user places his or her selected finger. Each terminal is connected to the central processor by two communication lines an RS-232 serial line for control information and an RS-170 video line for the live fingerprint image.

An optional surveillance camera may be fitted to overlook the portal location. In an alarm situation, such as someone tampering with the terminal, this camera can be switched automatically to give a live video display at the operator's console. An exit

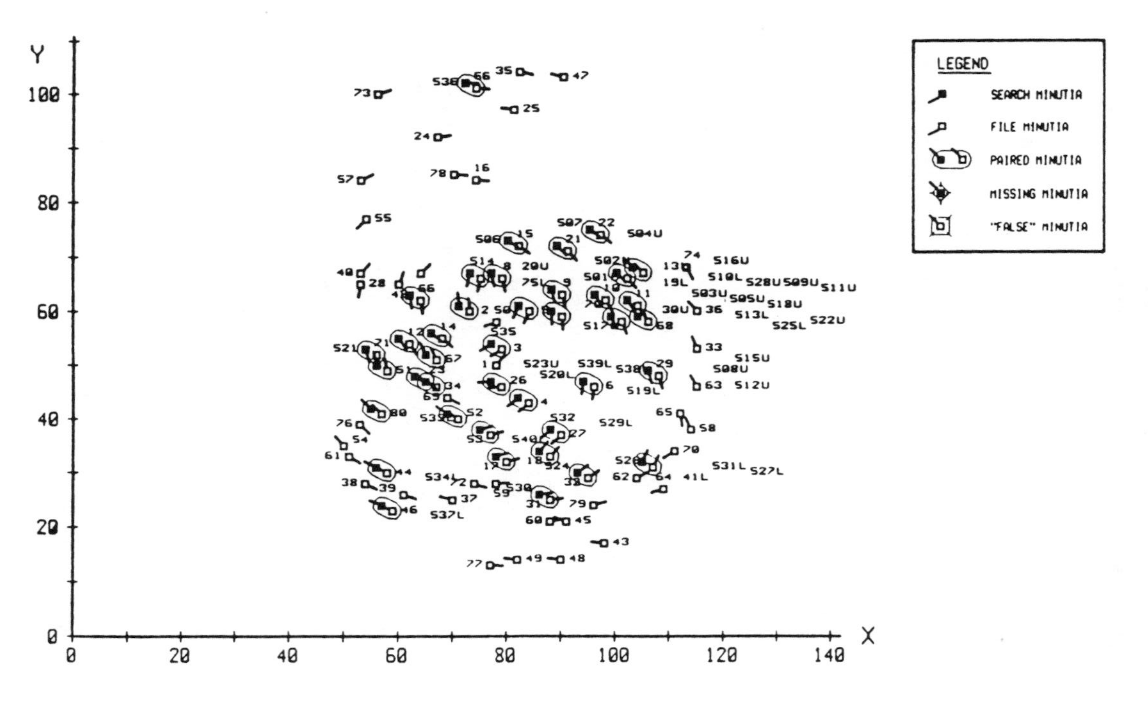

Figure 2. Minutiae matching.

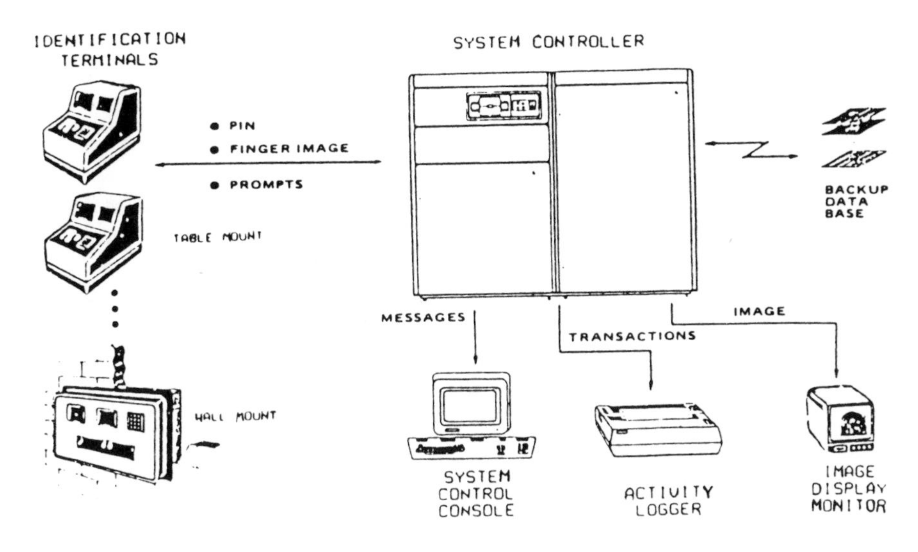

Figure 3. Components of PIV-100 system.

terminal may also be fitted inside the portal, allowing each user to identify himself when exiting from the protected area.

Figure 4 shows the architecture of the central processor. Serial lines from the terminals are handled by a DHV11 multiplexer, and the video lines by a conventional video selection switch, digitizer and frame buffer. Fingerprint images are enhanced and their minutiae extracted by our proprietary image processor, which occupies a separate cabinet of electronics. The KDJ-11/73 processor is supported by an MXV11 multi-function board and an MSV11 memory board giving a total of 640 Kbytes memory. The system database is stored on the standard RD51 10-Mbyte Winchester disk, backed up by the RX50 floppy disks.

OPERATION

Enrollment of a user into the database of the PIV-100 system is performed by the system operator and takes less than two minutes. First the user is assigned a four-digit personal identification number (PIN), either of his own choosing, or generated randomly. This number is checked against the active user directory to preclude duplication. Next the user's name and organization code are entered, and the appropriate authorization rights are defined. A user may belong to one or two groups, each of which may specify access by hour-of-day and day-of-week independently for each protected area. Individual user privileges may also be specified to override the group assignments, e.g., for overtime working or vacation periods. Finally, the user's "best" and "second-best" fingers, denoted as primary and alternate, are selected. Multiple scans of each

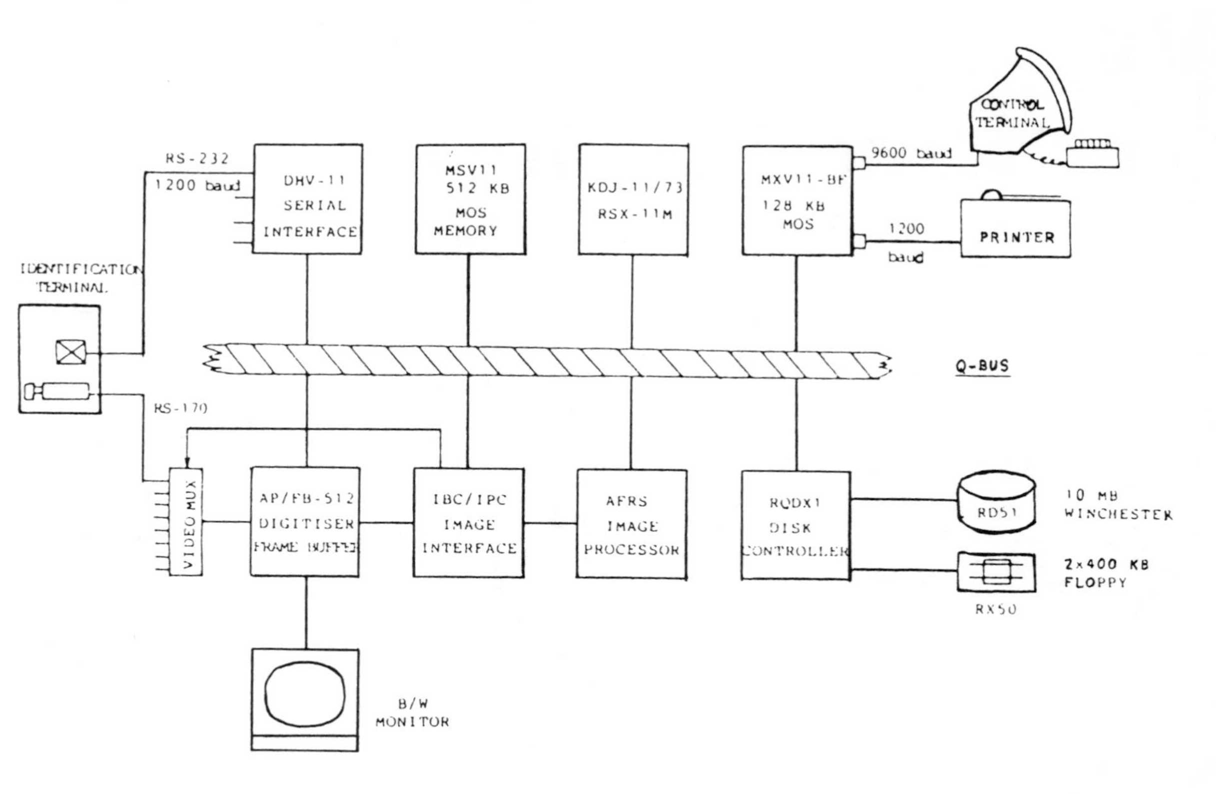

Figure 4. Architecture of central processor.

finger are made, in order to extract the most consistent set of minutiae for storage in the database.

The processing steps associated with an access request are shown in Figure 5. The cycle begins when a user comes to the terminal and keys in his or her PIN number. The system checks this number for validity, and then retrieves the user's enrollment record from the database. If the user is authorized for access into the protected area at the current time, he is asked to place his primary finger on the terminal's finger scanner. The fingerprint image is relayed to the central processor, its minutiae are extracted and then matched against those in the database record. If the two sets of minutiae match sufficiently well then the user is admitted to the protected area. The entire cycle typically takes about ten seconds, of which only two seconds is matching computation time.

What do we mean by matching the two sets of minutiae "sufficiently well"? The matching algorithm generates an integer score, which is a measure of the correlation of the two sets. The verification algorithm compares this score with an upper and lower threshold, initially set at the time of user enrollment. If the score exceeds the upper threshold, the user is considered to be verified. If the score falls below the lower threshold, the user is rejected. If the score falls in between the two thresholds then a retry procedure is initiated, as determined by the verification strategy in force for the protected area. A typical strategy is shown in flowchart form in Figure 6, with one retry attempt allowed for the primary finger, and a possible try and retry for an "alternate" finger.

The PIV-100 allows a separate verification strategy to be set for each protected area. In addition to the number of retries for each finger, it is possible to mandate

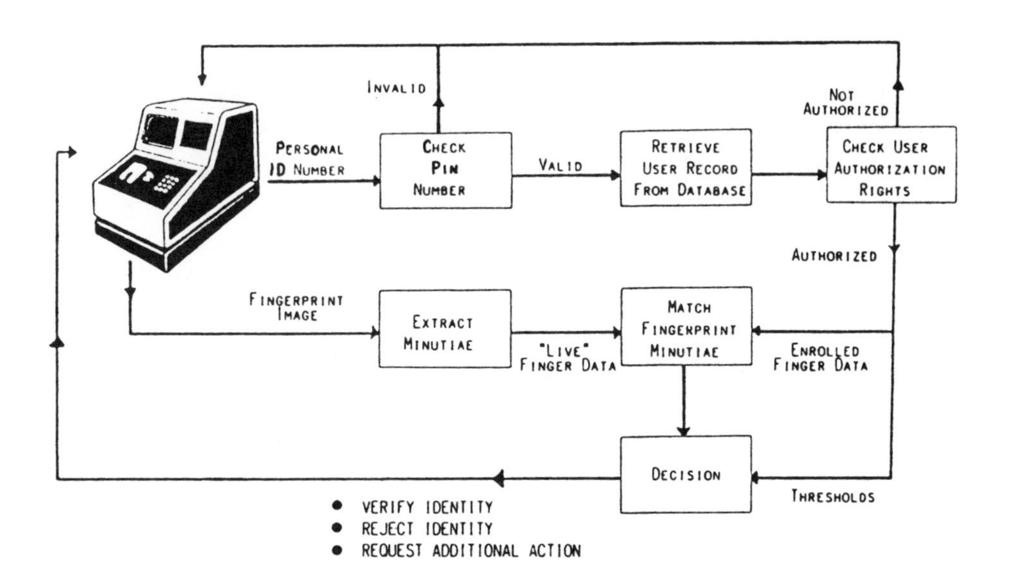

Figure 5. Access cycle processing.

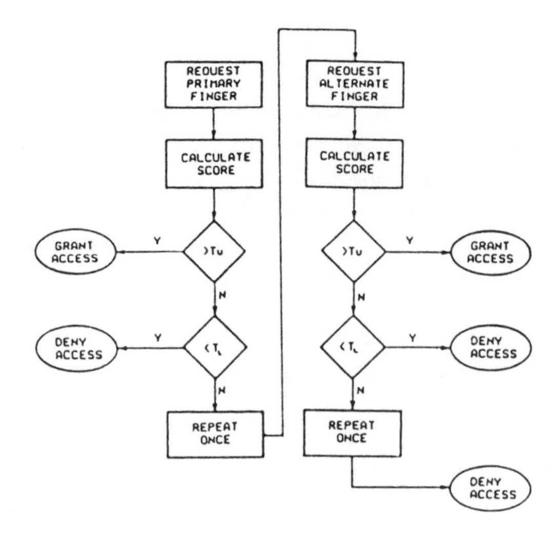

Figure 6. Typical verification strategy.

entry of the alternate finger for high-security applications, or to inhibit it when throughput rate is more important.

Error rates are related to the overall distribution of match scores.[15] If the probabilities of ''in-class'' and ''out-of-class'' scores are plotted against score value, we obtain the overlapping distributions shown in Figure 7. It can be seen that the optimum setting of a single decision threshold is critical, and that errors will occur in the region of overlap. With two decision thresholds, errors can be minimized by undertaking a retry procedure in the area of uncertainty.

In the PIV-100 system the actual error rates can be set to any specified level, by the adjustment of a number of inter-related system parameters such as decision thresh-

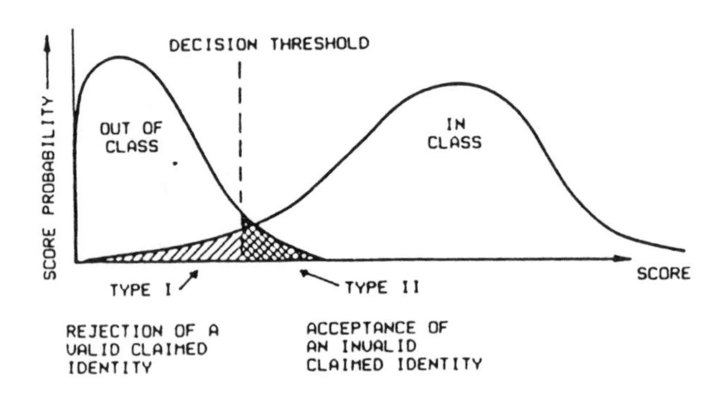

Figure 7. Score probability distributions.

olds, matching characteristics, and verification strategies. As always, one must trade off increased accuracy against reduced processing speed and convenience.

IMPLEMENTATION

In order to develop an effective design for the system, the Jackson System Development (JSD) method was used.[16] First the real-world entities of user and protected area were identified, and structure diagrams were drawn to show all the actions performed or suffered by each entity throughout the course of their respective life-cycles. The system was modelled to show all the inter-connections between the entity processes and the outside world, then elaborated to include all necessary functional outputs. Only when we had completely specified this model as a set of pseudo-code modules, did we begin to transform the design into an implementation in actual hardware and software.

The DEC Micro-11/73 was chosen as a suitable processor, with RSX-11M as a real-time multi-tasking executive. The entity pseudo-code was transformed into a group of tasks, communicating by event flags and data packets. These were programmed mainly in Fortran-77, with some use of Macro for time-critical functions. Extensive use was made of memory-resident tables and linked lists to represent the states of the various entities. The JSD technique of program inversion proved to be very valuable, allowing implementation of all the sequential terminal processes by a single scheduler, dispatching control to program segments under control of a ''text pointer'' variable maintained for each terminal.

The database design turned out to be very simple, a direct-access file of fixed-length records, with a memory-resident directory relating user PIN number to virtual block number. This directory is searched by a ''binary chop'' technique[17] for efficiency. Random PIN numbers are generated by a simple but extremely effective algorithm quoted by Ralston,[18] where the fifth and seventeenth most recent integers of a sequence are combined successively.

The operator interface, through the control/display terminal, is friendly, informative and tolerant to keying errors.[19] All display and control modes may be selected through a hierarchy of menu screens, with a menu at each level selected by a single key-stroke (1–9). Help screens are available at any time, corresponding to the current menu, and explanatory prompts for each data field are displayed in a four-line ''guide box'' common to every screen. The bottom line of the screen is reserved for a status display in reverse video, showing the time of day and the current number of users in each protected area. In alarm situations the guide box is overwritten by a bold reverse-video description of the alarm condition and advice to the operator on how to proceed.

CONCLUSIONS

In many cases, ''security'' can be defined as the control of access to a protected item or location, in such a way that only an authorized group of individuals may have such access. In order to achieve effective security control, each individual's identity must be verified by a personal attribute that is positive, accurate, reliable, easy to present, and not transferable. The human fingerprint meets all these criteria, and is the most widely accepted unique personal attribute, resulting from over one hundred years of research and usage throughout the world.

The De La Rue Printrak PIV-100 system, based on the matching of the minutiae extracted from a user's live fingerprint with those held in a database, represents the most advanced and reliable positive identification technology available today. It can serve as a stand-alone identity verification system, or as a component of a more elaborate security system. Reading directly from an individual's finger and completing its verification or rejection of identity in less than two seconds, the PIV-100 is suitable for use in virtually any application for high-security physical access control.

REFERENCES

1. T.S. Perry and P. Wallich, ''Can Computer Crime be Stopped?'' *IEEE Spectrum* vol. 21, no. 5, May 1984, pp. 34–45.
2. George H. Warfel, *Identification Technologies* (Springfield, IL: Thomas Books, 1979).
3. ''Guideline on User Authentication Techniques for Computer Network Access Control,'' U.S. National Bureau of Standards, FIPS Publication 83, September 1980.
4. ''Guidelines on Evaluation of Techniques for Automated Personal Identification,'' U.S. National Bureau of Standards, FIPS Publication 48, April 1977.
5. 'The Henry System,'' Identification Services Division, Royal Canadian Mounted Police (Ottawa), 1970.
6. ''The Science of Fingerprints Classification and Uses,'' U.S. Federal Bureau of Investigation (Washington), 1973.
7. B. Moayer and K.S. Fu, ''A Syntatic Approach to Fingerprint Pattern Recognition,'' *Pattern Recognition* vol. 7, 1975, pp. 1–23.
8. K. Rao and K. Balck, ''Type Classification of Fingerprints: A Syntactic Approach,'' *IEEE Trans on Pattern Analysis & Machine Intelligence*, vol. PAM1-2, no. 3, May 1980, pp. 223–31.
9. E. Hecht and A. Zajac, *Optics* (Reading, MA: Addison-Wesley, 1975), pp. 81–4.
10. David G. Edwards, ''Fingerprint Sensor,'' (Siemens Corp), US Patent 4,429,413. January 1984.
11. K.C. Petersen, ''Silicon as a Mechanical Material,'' Proceedings, IEEE vol. 70, no. 5, May 1982, pp. 420–57.
12. Joseph H. Wegstein, ''The M40 Fingerprint Matcher,'' U.S. National Bureau of Standards Technical Note 878, July 1975.
13. D.M. Osgard and S.A. Smithson, ''Automated Fingerprint Identity Verification,'' Proceedings, IEEE Conf. Electro. '77, New York, April 1977.
14. ''Entry-Control Systems Handbook,'' Sandia Laboratories SAND80-1033 (Albuquerque, NM), September 1980.

15. ''False Identification, the Problem and Technical Options,'' U.S. Dept. of Health & Human Services (Washington), 1983.
16. Michael A. Jackson, *System Development* (London: Prentice-Hall International, 1983).
17. Donald E. Knuth, *Sorting and Searching*: The Art of Computer Programming, vol. 3 (Reading MA: Addison-Wesley, 1975), p. 406–7.
18. Anthony Ralston, *Encyclopedia of Computer Science & Engineering*, 2nd ed. (New York: Van Nostrand Reinhold, 1983), p. 1264.
19. Norman Luckett, ''An Efficient Interactive Menu Handler for VT-100/200 Displays,'' presentation to Graphic Applications SIG, DECUS U.S. Symposium, Anaheim, December 1984.

PART VIII

Unique Security Applications of Technology

The previous articles dealt with conventional security solutions. Albeit, they are modern devices and sophisticated systems. This section discusses highly unique and unconventional solutions.

The first article in this section discusses security robotics. Robotics is a well established discipline. Robots have proven their value in industry, replacing humans in repetitious and boring jobs, performing their tasks better and cheaper, leaving the more complicated and challenging jobs for people. When one considers the high cost of security personnel, the often repetitive and boring nature of many of their tasks, it is logical for administrators to think of robots for security.

The first article is a rather general survey of the current state of security robotics. It addresses the question "Can security robots replace humans?" The answer depends on the specific problems. [The first article assists security administrators in determining the problems associated with security robots and where the appropriate robots can be purchased.]

The second article deals with old security solutions using animals, but highlights their modern-day applications. Some of these sophisticated applications are enhanced by advanced knowledge of animal behavior, such as using animals in earthquake prediction. Other modern uses are made possible by sophisticated interfacing of electronics and animals, for example, Dynatrend's use of dogs in perimeter intrusion detection.

The use of animals in modern security applications is often merely a rediscovery of ancient uses, exemplified in the use of Hiram Walker distilleries sentry geese. Following this example, the U.S. Army in Germany is now experimenting with sentry geese.

One usually thinks of applied physics, electronics or mechanics in connection with the term *technology*. Technology is really the application of science to practical problems. The application of animal science to security problems is most certainly technology. Matters of semantics aside, the real issue is obtaining the most effective, economical and aesthetic solution consistent with the image of the organization. The knowledge of the application of animals to security problems appropriately expands the security director's repertoire of solutions.

The final article in this section outlines a solution to the growing problem of emergency monitoring systems for the elderly. With the increased number of elderly people and the concomitant sky rocketing cost of medical services, the need for effective and economical care for the elderly is reaching crisis proportions. There is not enough space in rest homes to house those who require care, and costs are great for those who are admitted. Hospitals are also forced to release those who do not need hospital care. The pressure, then, is for more elderly to remain in their own dwellings. Security technology may be able to greatly facilitate this by enhancing their security and safety and providing immediate emergency service when necessary for those living at home.

Green's article demonstrates an excellent security solution to this most grievous problem. It explicitly outlines the parameters of the problem and qualitatively informs us as to how modern technology addresses each parameter to form an excellent and cost-effective solution to the problem. The beneficiaries of these solutions are the vast aggregate of the American elderly and the nation as a whole.

Green explicitly describes a system using state of the art technology, especially computer technology. He also gives us, albeit implicitly, insight into the problem in the identification and planning process involved in solving unique security problems.

Security Robots

Romine (Dick) Deming

Abstract. It has been speculated that robots will invade the security world, saving human lives and replacing costly personnel. Robots have successfully made a place for themselves in industry, accomplishing their tasks cheaper and better, in most cases, than people. They could be used for the security functions of patrol, performing routine mechanical tasks, neutralizing dangerous situations and as responders to general threats. The technology certainly exists. However, a number of thwarts limit the stream of robots into security. Not the least is a preference for familiar security solutions especially the use of security officers.

POTENTIAL USES

"The robots are coming! The robots are coming!"[1] so states a 1983 Security Management article. Or, are they? Security robots do offer exotic solutions to many security problems. The potential most often thought of is in patrol or watchman work. Robots could patrol large complexes, and do it in total darkness, saving on electricity costs. Of course, they could be equipped with multiple sensors. It is arguable that the "bad guys" cannot violate a unit which changes its position and is not attached to any standard system. And, robots will work for practically nothing.

There are other security problems responsive to robots. There are a number of routine mechanical operations performed by some security officers such as opening and closing doors and heavy gates. These functions can be accomplished much cheaper by robots and with equal or better facility.

Also, robots are expendable where humans are not. An explosion to a robot merely provides salvage. Therefore, robots are valuable in neutralizing life-threatening situations such as bomb detection and disposal.

Finally, robots have potential as general responders as the following scenario illustrates: a nuclear power plant is located sixty miles up a mountain. A single security

1985 Carnahan Conference on Security Technology, University of Kentucky, Lexington, Kentucky, May 15–17, 1985.

officer is on duty and a few maintenance engineers. The perimeter sensors indicate a breach attempt has occurred in sector 4. Three robots automatically respond. Upon arrival at the breach, they confront the trespassers and detect they are armed. Robots Alpha and Beta fire tranquilizers into the trespassers. Omega then carries them to a detention area. Alpha and Beta remain at the breach for the evening.

AVAILABLE ROBOTIC TECHNOLOGY

Are these potentialities flights of fancy for security directors or are they realistic possibilities? Industrial robots are already successfully doing similar tasks and even many more than those discussed above. $240 million was spent on industrial robots in 1983.[2] The 1981 robot census indicated some 4,700 robotic systems in operation. Robots are multiplying at a rate of 30 percent per year. The World Future Society predicts 35,000 robots will be installed by 1990.[3]

Robots are used to perform precise and intricate welding and soldering. Bin sorting robots can make even fine discriminations. Mobile base robots pick up and convey parts to assembling robots. Voice synthesizers are available as well as conventional sensors.

Robots are being developed to save lives. Carnegie-Mellon University researchers are working on robots to build mine roofs, one of the most dangerous jobs in coal mining. Another project of this University is developing a snakelike robot to dig up natural-gas pipes, then blow the dangerous fumes away before humans enter the area. This work now claims 150 lives a year.[4]

WHAT IS A ROBOT?

A robot is described by the definition of The Robot Institute of America. "A robot is a reprogrammable multifunctional manipulator designed to move material, parts, tools or specialized devices, through variable programmed motions for the performance of a variety of tasks."[5]

The key words in the definition are "reprogrammable multifunctional manipulator." Achieving this behavior has made robots the important alternatives to production personnel. Computer development has made robots possible. The same printed circuit technology which has enhanced computer development has greatly reduced the size of electrical components, making mobile base robots a reality. Robots are capable of small mass equipped with on board micro processers, numerous sensors and multiple motors for movement and manipulation. AI (artificial intelligence) permits robots to learn from mistakes. The improvement in sensing technology has produced robots with highly discriminatory powers.[6]

ROBOTS IN SECURITY

All of this technology developed for industrial robots is applicable to security as well. What is the state of the art in security robotics? There are a variety of security robots available and a relatively small cadre on duty.

Robots on Patrol

Patrol robots are on the market. Odetics Inc. produces ODEX 1 for $200,000. This figure constitutes a payback of one and a half years if replacing officers for one post. The six-legged unit resembling a daddy longlegs can traverse almost any terrain carrying up to 1,800 lbs. Its dome head houses a TV camera.[7]

A mechanically inclined security director can learn about robotics while assembling his or her own robot for a cost of about $2,000. When assembled, the Hero 1 will be completely self-powered, capable of sensing light, sound and motion, with a gripper arm and speech synthesizer.[8]

There is a report of a security robot built by a group of engineers in their spare time. Whether this report is fact or fiction is unknown. However, the technology is readily available, which lends credibility to the report.[9]

The robot, Joey, is a cute but functional hemisphere on wheels. His primary function is locating dangerous sources of heat in the electronics laboratory located atop a 12,000 foot mountain. The distance from town prevents timely assistance by public firefighters. To accomplish his mission Joey is equipped with four sensors: infrared, a wide angle lens photo cell to see room illumination, photosensors to follow pilot lamps that mark the electrical outlets on the work benches and a low power sensor. This last sensor directs Joey to go to his power source for a battery exchange automatically. His infrared sensor is connected to an on board audio alarm. Joey spends his time traveling throughout the laboratory looking for unusual heat sources which might indicate a fire.

Although the technology exists for patrol robots, the economic viability is questionable for most applications. If comparing the cost of patrol personnel, there is no doubt the patrol robots are most cost effective in the long run. But, comparing a mobile base ODEX 1 or even Hero 1 against stationary sensors such as infrared, microwave, buried line, vibration sensors, pressure mats, or CCTV with motion or alarm activation, the cost becomes a real concern.

Performing Mechanical Operations

One can question whether gate and door openers are true robots. If interfaced with sensors and micro processors which permit reprogrammable discrimination and access, then it is arguable that they are indeed robots. Counted as robots then, this type has achieved a place in many facilities. At least four companies are extensive producers of stationary base robots employed as gate and door tenders.[10]

Most access control systems can be interfaced to replace "turn key" personnel in the boring job of tending gates and doors. In fact, a robot interfaced with an Eyedentifyer™ can do it better.

An unauthorized individual using a stolen ID badge may disguise him- or herself to look like the authorized person on the badge to fool an officer. However, the robot cannot be fooled. Imposters cannot change the blood vessel configuration of their retinas for the robot to open the door, no matter how persuasive they become.

There may be other mechanical tasks for security robots. They could be used for checking public passengers and their luggage for weapons. They could also examine mail for explosives.

Neutralizing Dangerous Situations

Human lives are precious. At least two companies produce robots for neutralizing dangerous situations such as bombs or armed confrontations. Morfax Limited has been selling the Wheelbarrow MK 7, an innocuous enough label, for eight years. There are more than 300 in use in more than 25 different countries. It is a mobile-based robot designed to investigate, neutralize and remove hazardous devices and life-threatening people. It can carry a CCTV or X-ray camera. It will fire automatic weapons including tear gas. It can ascend stairs and maneuver in tight places.[11]

Ion Track Instruments, Inc., produces a similar robot for the same missions with rapid change wheels/tracks.[12] Its unit is generator driven rather than battery driven as is the Wheelbarrow.

General Threat Responders

Although no known robots exist as general responders to human threats, the advantages seem abundantly clear. A facility may have the best security sensors available and they may be well integrated into layered systems. If no one is there to respond to an attempted breach, of what value is it to know of the attack? Yet, it is extremely costly to have a response team available and trained to the peak of readiness. By the same token, to wait months and years for a scramble which hopefully will never occur is extremely boring. And when the attack does occur, it may be dangerous; human life may be needlessly risked.

Security robot responders could be designed to respond as discussed earlier. Mobile robots perform a number of similar functions in industry.

Thwarts to Security Robots

Robots are barred from security service by a number of interrelated factors. The lack of the necessary knowledge definitely is not one of those factors. We have the know how.

One of the primary factors thwarting security robots is competition from more economical alternatives. A company can install a lot of buried line for the cost of a sufficient number of robots for equal coverage. Volumetric motion detectors are competitive to the necessary patrol robots. There may be exceptions such as highly mobile items to secure as in ammunition dumps, motor pools or construction sites.

This brings us to another factor. Although a need for a security robot may exist, the need for the specific type of robot may be so unique that it cannot justify research, development and production. Economy of scale will remain an important variable in industry and commerce. Small-volume sales are not popular.

Problem identification is another factor. Someone has to identify the problems responsive to robotic technology. Those involved, it appears, will be kept busy in the industrial field for many years to come. There seems to be little attraction to security problems for robotic engineers as there was in the electronics field when the 1970s lull in the space industry occurred.

The greatest factor thwarting security robots appears to be the orientation of users or security directors. Security directors hale from the military or police organizations or from professional educations, oriented to human solutions to security problems. Some administrators are becoming avid technocrats. Others are acquiescing to technology. However, at least according to security system design planners, manufacturers and venders, a technological orientation in the security field is minuscule. Before the robots can come to security the users will have to become more robot friendly.

CONCLUSIONS

Are robots coming to security? The potential benefits are great. The necessary technical knowledge exists to produce a wide range of security robots to cost effectively replace budget decimating personnel to solve problems too costly or dangerous for people. They could remove much of the drudgery of a security officer's work.

Yes, robots are here as a token force. However there are many impediments to the aid of harried directors with gargantuan budgets.

REFERENCES

1. William Cole Cathey, ''The Robots are Coming! The Robots,'' *Security Management* 17 (September 1983), p. 113.
2. John G. Fuller, ''Death by Robot,'' *OMNI*, 6 (March 1984), p. 45.
3. Harry F. Rosenthal, ''Bottom line to forecast by futurists: 'Enjoy today,' '' *Associated Press* (December 26, 1984).

4. Fuller, ''Death by Robot,'' p. 45
5. Robot Institute of America, *Worldwide Robotics Survey and Directory*, P.O. Box 1366, Dearborn, Michigan 48121.
6. William J. Higgins, ''Robots in Manufacturing,'' *Machine and Tool Blue Book*, 79 (June 1984), p. 46.
7. Odetics Inc., Anaheim, California.
8. NRI Schools, McGraw-Hill Continuing Education Center, 3939 Wisconsin Ave., Washington, D.C. 20016. Kits are also marketed through 65 Heath-Zenith stores throughout the U.S.A. at a cost of $600 to $850.
9. E. Sorensen Jesby, ''How Did Joey Die?'' *Technology Illustrated* (April/May 1982), p. 82.
10. Robot Industries, Inc., 7041 Orchard, Dearborn, Michigan 48126. Pitts Security Gates Ltd., Bonehurst Road, Horley, Surrey RH6 8PP, England. Richards-Wilcox, 788 Third St., Aurora, Illinois 66507. Stanley Automotive Openers, 5740 E. Nevada, Detroit, Michigan 48234.
11. Morfax Limited, Willow Lane, Mitcham, Surrey CR4 4TD, England.
12. Ion Track Instruments, Inc., 109 Terrace Hall Ave., Burlington, MA 01803.

Modern Security Uses of Animals

Romine (Dick) Deming

Abstract. Advances in security applications of high technology may be eclipsing modern uses of animals. Animals have been historically valuable security resources. They may have unique benefits for security problems not responsive to high technology or not cost effective for the particular problem. Some of those problems may be responsive to animal use alone or in interface with technology. Some innovative security uses include Hearing Ear Dogs, Dynatrend ALIAS™ System, Hiram Walker Sentry Geese and China's Earthquake Warning System.

INTRODUCTION

As we continue advancing into greater reliance on high technology in security and safety, we may be overlooking modern applications of previous solutions, the use of animals. This is especially dysfunctional in situations where problems are unresponsive to high technology or where the solutions are not cost effective. Animals have always been and will continue to be valuable allies in human security and safety.

History tells us the Romans developed barking dogs from wolves (as a species, wolves do not bark). Alexander the Great used guinea fowl to protect his camps from night attack. Treasures have been protected by venomous reptiles. Peasants have successfully predicted natural disasters such as fires, floods and earthquakes from animal behavior. What modern uses exist?

EXAMPLES OF MODERN SECURITY USES OF ANIMALS

After a comprehensive literature search, we found a dirth of published information on modern security uses of animals. However, we have learned of some excellent uses

1984 Carnahan Conference on Security Technology, University of Kentucky, Lexington, Kentucky, May 16–18, 1984.

from word of mouth. We are sharing some of the more intriguing published examples and recording in professional literature some of the uses we have visited in the field.

The Security Value of Dogs

When one thinks of security uses of animals, one tends naturally to think of the guard dog or attack dog, and the infamous junk yard dog. More recently dogs have been successfully used in sniffing out explosives. The potential for lifesaving in this area is great when one thinks of the terrorist car bombing at London's Harrod's Department Store on December 17, 1983 killing six people and injuring 94.

Of course, dogs are used extensively for security incidental to other functions. It is a well-recognized fact that the family pet, in addition to providing love and companionship, is also an effective thwart to home burglars. Seeing Eye Dogs are in common use, providing guidance and companionship to the blind. In addition they protect their masters and mistresses from dangers such as speeding cars, misplaced objects around the home, fires, and the street mugger looking for an easy mark.

Hearing Ear Dogs

Hearing ear dogs are a new and innovative use of dogs with security implications.[1] The Hearing Ear Dog Program reported here is the oldest of fourteen such programs nationwide. Only three years old, it is located in the Worcester, Massachusetts area. It serves a small number of the twelve million deaf people in America at a cost of only $150 to the deaf person. (The remainder of the $1800 is donated by such charitable organizations as the Grange, Kiwanis and Lions). In addition to deterring burglars and muggers (which any family house dog would do), the dogs also provide specific services to their deaf owners, some of which are security and safety services.

The Program obtains dog candidates between the ages of eight months and two years from dog pounds, humane societies, and breeders to match with client applicants. The dogs are then given three to four months of training in socialization, obedience, and sound. At the time of application, clients select six sounds for their dogs. Standard sounds include a baby crying, smoke alarm, door knock, door bell, the person's name, telephone, cars, siren, keys dropping, tea kettle, and oven timer. When the dogs are sufficiently trained, clients come to the ninety acre Hearing Dog Program facility for a two-week stay. Under the tutelage of a coach, they learn to care for and work with their future canine partner. Most important, the clients are taught how to continue to train their dogs with other sounds.

Sound training consists of teaching the dogs to respond to the sound by leading its owner to the source with the exceptions of warning the master of dangerous sounds and in the case of a smoke alarm when the dog leads the master to the nearest safe exit.

The security implications of hearing ear dogs are obvious. The dog helps a deaf parent protect his or her baby. A special case is the deaf parents whose baby is subject to the Infant Sudden Death Syndrome. In addition to leading their owners to safety in response to a smoke alarm, dogs can be trained to respond to burglar alarms by ferocious barking. A unique example is the hearing ear dog which is especially valuable to a gentleman from Texas whose work requires him to be in areas inhabited by rattlesnakes. His dog gives him the security to go about his work relatively free of anxiety, knowing his hearing ear dog will warn him of the presence of any diamondback rattlesnake.

The dogs provide an additional, if not more subtle security function to their masters. People who lose their hearing often experience a serious sense of alienation or loneliness. This can produce fear and anxiety, depression, and a response similar to paranoia. The Hearing Ear Dogs, both by their companionship and the hearing service, give their masters confidence in their own ability to successfully deal with their environments. In at least one case, the Hearing Ear Dog is credited in preventing the death of a person who was suicidal prior to receiving his Hearing Ear Dog.

Dynatrend Interfaces Dogs with Electronics for Intrusion Detection

Dynatrend Inc. of Woburn, Massachusetts has successfully interfaced dogs with electronic devices and microprocessors to economically secure large areas containing critical contents (reported by Ronald Lawson at the 1983 Carnahan Conference).[2] The Adaptive, Large-area, Intrusion Alarm System (ALIAS™) employs trained dogs to patrol large areas such as fenced external areas including weapons, supplies, and equipment storage areas, and aircraft parking areas. The system can be adapted to secure enclosed spaces such as warehouses, manufacturing plants, and aircraft hangers. It is also adaptable to large unfenced tracts such as manufacturing and office complexes, and ballistic test sites and construction sites.

Some unique benefits of ALIAS™ is its flexibility, ease of installation, and effectiveness during periods of low visibility due to rain, snow, and fog. The system is readily adaptable to variations in quantities and localities of items to be protected without extensive hardware or software modifications.

The natural roaming instincts of dogs are used to detect hazards. They are trained to roam on ultrasonic command, identify threats and appropriately respond after identification. The mere presence of the dogs provides some degree of natural deterrence. However, the dogs respond to specific threats by going to a pressure pad upon encountering a threat, thus communicating that a threat has been sensed. The response is transmitted from the pressure mat to the central processor. The processor then emits a graduated series of messages or an instant message to an appropriate alarm annunciator. When the alarm situation has been neutralized, the dog is given an ultrasonic command to resume wandering in its sector. Capabilities exist for automatic self-test of the system, and a duress signal from the dog in case it is immobilized.

Hiram Walker's Scotch Watch Sentry Geese

Although dogs are perhaps the most common security animals, geese have also enjoyed a long and famous reputation. Sophisticated security design calls for achieving the best security solution at the lowest cost, which is aesthetically pleasing and consistent with the image of the organization. Hiram Walker Distilleries Limited has achieved this balance by using White China Geese to protect its coveted whiskey.[3] Their bonded whiskey must be stored in bonds (warehouses) for long periods of time. People were known to break in and help themselves to this aging nectar. The company started using geese nineteen years ago and has solved its security problem.

In addition, their geese could produce a profit. However, Hiram Walker chooses to sell the eggs and goslings to employees and donate the money to charity. The geese cost little to raise because they are grazers and eat the grass of their compounds. Occasionally they are given mash as a treat, a by product of the distilling process.

The geese are consistent with the image the company wants to convey. In fact, the geese are incorporated in the advertising "Ballantine Bonded Whiskey, Protected for You by the Ballantine's Scotch Watch Geese" (Ballantine Scotch is a product of the Hiram Walker Company).

The aesthetic value of the geese is obvious to travelers passing by Hiram Walker distilleries and bonds in Scotland and Canada. One is struck by the beauty which these large graceful white birds lend to the facilities they guard as they lounge in the lush green compounds by their aqua ponds.

The facility I had the pleasure of seeing is the Ballantine Bonds located just south of Loch Lomond on the Firth of Clyde near the town of Dumbarton in Scotland. The perimeter of the facility has a double fence barrier approximately 100 feet between the fences. The area is then sectioned off, producing a number of large separate pens. Each pen contains a pond and a flock of geese. The total desired complement is 200.

Now, all geese have loud voices. However, the White China geese are known for their noisiness. Unlike their heavy, more sedentary cousins, Embden and Toulouse geese, the Chinas are very sleek, agile, and active—well suited for the role of sentries. Some of the flock in each pen will always be on duty while the others eat, sleep, or copulate, even at night. When a thirsty, would be trespasser approaches a pen from the outside, the sentries on duty in the pen alert the others. The others then alert the geese in the additional pens and immediately 200 geese are announcing the presence of a trespasser. Most people are aware of the potential noise of geese. If this is not enough to deter a thirsty thief, then the exposure to the sonic blast of two hundred screaming geese probably will. However, there are personnel on duty to respond in case the culprit is really thirsty.

Animals as China's Earthquake Warning System

Security from earthquakes is definitely important because of the potential for extensive harm to a great number of people and property. The 1976 quake of Tangshan, China took 655,000 lives. A previous quake with an even greater potential for damage claimed few victims because of the use of animals as part of an earthquake warning system. Peasants of areas with a high incidence of earthquakes have known of the unusual animal behavior to warn of an impending earthquake for a long time. In many of these areas the folk information has been used to prevent loss of life and property. However, this information has been informal and unsystematically applied in the past. China offers an excellent example of modern systematic use of animal behavior to warn of impending killer earthquakes.

Unlike the other examples reported here, the use of animals for earthquake warning does not require training animals, but training people to be alert to the unique natural behavior of animals prior to earthquakes. This training is justified because animals are the only viable alarm system for short-term earthquake predictions.

Tributsch, after careful scholarship, found more than 100 statements recorded in written history of unusual animal behavior just prior to and predictive of a serious earthquake from all over the world. The first account goes back to 373 BC in Helice, Greece. This reported number is the more remarkable because most observations are by peasants who have intimate contact with animals. Yet peasants are not likely to communicate their observations in scientific journals. Since the late 1970s, a few researchers have interviewed peasants after quakes and have reported their findings adding to the number of the historical accounts. Funds for this type of research are scarce, however, constricting the quantity of this type of data.[4]

Both historical and contemporary data from all over the world are consistent regarding the response of animals to impending quakes. Even prior to a seismic reading of a quake, animals begin to show signs of severe agitation or panic. The length of the warning time varies with the size of the quake, the climatic conditions, geological factors, and the distance between the alarming animals and the epicenter of the quake. The longest advanced warning recorded is 21 hours. Most typically, the alarm is within an hour or less of the quake. The more common responses of animals are: Horses balking and refusing to move forward, throwing themselves down on the ground, or stampeding toward the direction from which they traveled. Dogs howl in unison. Dogs and cats try madly to leave the confines of buildings. Poultry refuse to go into their shelters. Wild rodents migrate in mass. Wild animals lose their fear of man. Flocks of birds fly at unusually high altitudes. Bees swarm from their hives. Reptiles leave their holes, even during hibernation. Lastly, fish throw themselves out of the water and often on to the shore.

No one has really determined the stimuli for these responses. These are a number of hypotheses: weak preceeding vibrations, unique electrical fields, unique magnetic fields, unique gaseous odors, and low levels of electromagnetic radiation. The last seems to be the most convincing as the stimulus. This hypothesis states that extensive pressure exerted against rock, especially rock containing large deposits of quartz, emits low levels of electromagnetic energy. This energy is picked up on the sensitive hairs, skin, or scales of the animals. Because of the unusualness of the stimulus, the animals have no standard response. They do not know what is likely to happen, so they panic. They wish to leave the perceived source of the stimulus. It is believed that house animals wish to escape from enclosed areas because the energy affixes to dust particles. These particles are more concentrated in enclosed areas giving animals a higher dose of electromagnetic energy. The same situation occurs in chicken coops. Chickens wish to leave or refuse to enter their shelters because of the higher energy content from electromagnetic energy affixing to chicken dust.[5]

It makes relatively little difference to us which stimulus produces the response. The fact remains that animals do respond in an obvious and unique way, common to all their species, prior to a serious earthquake, wherever it occurs. These responses can be and have been used to warn people of killer earthquakes.

The example from China illustrates the potential of organizing and training people to use the natural responses of animals to warn potential victims of impending killer earthquakes. But first it must be made clear that presently there is no scientific or technological alternative. As Frank Press of MIT pointed out at the 5th Japanese-American Seminar on Earthquake Predictions in 1977, the prospects of observing enough earthquakes experimentally in the next ten years to let scientists make reasonable valuable predictions based on collected information are very poor. The alternative is to use animals we know will respond, and communicate the nature of these responses so masses of people in earthquake areas can use them beneficially.[6]

The example of organized, formal, and successful use of animals to sufficientiy warn people so that they can protect themselves occurred in the Haicheng area of China on February 4, 1975. On that day at 7:36 P.M. in this heavily populated area, an earthquake of 7.3 on the Richter Scale struck. The area hardest hit was around the epicenter near Haicheng containing about half a million residents. Fifty percent (50%) of the buildings were destroyed, but there were few victims, only those who refused to be evacuated or respond.

The case begins in June of 1974. Following extensive geophysical study, China's National Earthquake Bureau warned that a serious earthquake would strike the Haicheng area within two years. As a result, the people were mobilized. Guided by experts, amateur groups were organized and trained by the thousands. They were from the industrial plants, schools, animal breeding institutions, and agricultural communes. They were trained in recognizing animal signs and reporting this information to a central clearing station. The 100,000 volunteers were constantly reminded of the importance of their work by flyers, lectures, personal instruction, school lessons, and radio broadcasts.[7]

In December 1974, the efforts began to bear fruit. From the reported animal responses, the Earthquake Bureau predicted a small earthquake would soon occur in the region where the large one was expected. It struck as predicted on December 22, 45 miles north of Haicheng. Preparations were intensified for the evacuation and care of people and to minimize property destruction. During January 1975 reports of unusual animal behavior intensified. A sudden surge occurred with the beginning of February. During the first days of February a number of small quakes were experienced and then decreased by the 4th day. However, by 10:00 A.M. on the 4th day the warning signs were unmistakably clear. Everything was ready because the experts were certain an earthquake was eminent. By 2:00 P.M. evacuations had nicely begun. Animals were moved into the open. Vehicles were parked away from buildings. And, people were moved to areas previously designated as shelters and into open areas. Movies were shown to keep their minds off the impending crisis. At 7:36 P.M. the quake struck as previously predicted.

The Chinese seismologists involved stated the observations of animals was the best method for earthquake prediction thus far.

The United States is taking cognizance of the potential of animal warnings. The U.S. Geological Survey has contracted with the Stanford Research Institute to establish a network of observers along the San Andreas fault. There is a central office receiving telephone reports from a few hundred selected collaborators observing the behavior of some 70 species of animals. I am confident we all wish them complete success in case a killer quake strikes that region.[8]

CONCLUSIONS

Examples of modern security uses of animals reported here seem to offer one or more important reasons for using animals. Animals may be the only alternative, such as in the case of earthquake warning. Perhaps they are the most viable as in the case of dogs and explosive device detection. They may fulfill a variety of other needs concomitant to their security function, such as Hearing Ear Dogs. They may be more cost effective than alternatives, as in the case of the Dynatrend System. Or, they may offer a number of other fringe benefits such as the Hiram Walker geese.

If we are sensitive to the potential of animals, we may discover yet unknown situations to successfully and cost effectively solve knotty security problems by animals either alone or in interface with high technology.

REFERENCES

1. Donald MacMunn, Hearing Ear Dog Program, 76 Bryant Road, Holden-Jefferson, MA 01522.
2. Ronald N. Lawson, Manager, Physical Security Applications, Dynatrend Inc., 21 Cabot Road, Woburn, MA 01801.
3. Hector MacLennan, Hiram Walker and Son (Distilleries), Ltd., Glasgow, Scotland.
4. Helmut Tributsch, *When the Snakes Awake* (Cambridge, Massachusetts: The MIT Press, 1982).
5. Ibid.
6. E.R. Lapwood, *Nature* 266:220 (1977).
7. C.B. Raleigh et al., "The Predictions of the Haicheng Earthquake," reported by Liaoning Earthquake Study Delegation, U.S.A., 1976.
8. L.S. Otis, "Project Earthquake Watch," Office of Earthquake Studies, MS 77, 345 Middlefield Road, Menlo Park, CA 94025.

Emergency Monitoring Systems for the Elderly

Peter E. Green

Abstract. This article describes the progress made in a research project to develop monitoring and emergency help-call systems for the elderly. The purpose of these systems is to enable the elderly to live autonomous lives in their own homes as an alternative to being institutionalized when they become infirm.

The paper reviews 20 years of field experience in England with help-call systems for the elderly. Then it describes a new microprocessor based help-call system that is being developed specifically for application on a city-wide basis in the USA. This system features a high level of security for inner city use and the use of two way voice mixed with DTMF tone signalling between the elder and a dispatcher at a help-call center.

The paper also describes current research into the use of artificial intelligence techniques combined with low-cost sensors for detecting the movement of elderly people.

INTRODUCTION

Twenty years ago, the author was the technical director of a team which developed the first elderly persons' emergency help-call system that was approved by the British Ministry of Health. Since then he has had the opportunity to follow the progress of these systems as a consultant to KRS Electronics Ltd. which has installed a large number of these systems in England. The author is a member of the faculty of Worcester Polytechnic Institute where he has recently initiated a research program into the application of advanced computer, sensing, and artificial intelligence techniques to non-contact monitoring of the elderly and handicapped. He is also technical director of a

1985 Carnahan Conference on Security Technology, University of Kentucky, Lexington, Kentucky, May 15–17, 1985.

team that is developing a new microprocessor based help-call system for Help Call Systems Corporation (HCS).

In Section 2 of this paper, the author describes the systems that are extensively used in elderly persons' apartment buildings in England. He then identifies the major issues that have been learned from this field experience. In section 3 the author describes systems in use in America and finds that they are functionally less advanced than the systems in use in England. In section 4 the author describes the new system that is under development by HCS which is functionally more advanced than the English systems and which is designed for use in an American environment. Finally, in section 5 the author describes the research program at WPI.

THE ENGLISH EXPERIENCE

The British Government has built a large number of elderly persons' apartment buildings. Almost all of these buildings are equipped with a help-call system that are more advanced versions of the same functional design developed 20 years ago. These elderly persons' apartment buildings are provided as an alternative to institutionalization in a nursing home for those elderly who, while suffering from some infirmity, are able to live autonomous lives. These apartment buildings typically house between 30 and 60 elders with each elder having the equivalent of a studio apartment with its own kitchen and bathroom. Usually these buildings are low-rise having 2, or at most 3, stories and have a common room where the elderly can socialize. The well-being of the elderly residents is supervised by a custodian who is typically a nurse. This custodian lives in one of the apartments and is available to help the elderly 24 hours a day.

These apartment buildings have been very successful and enable the residents not to be institutionalized except in case of medical emergency. A significant element in success is the provision of the help-call system which enables the elderly to call the custodian for help whenever they are in need. The custodian then organizes the help needed. The emergencies can be medical (in the case of a fall), environmental (such as an overflowing toilet), or emotional (such as loneliness).

A typical English help call system is shown in Figure 1. The elder's apartment is equipped with a number of pull-cord switches which can be pulled to summon help. These pull-cords are typically located by the bath, toilet, in the kitchen, by the bed, and by their favorite TV viewing chair. When the elder pulls on one of the cords, an alarm is sounded in the custodian's apartment. An indicator lamp is also lit on the custodian's control panel to indicate which elder needed help. The custodian establishes communication with the elder's apartment by pressing the appropriate selector switch and is then able to talk to the elder over the intercom system to determine what help is needed. Where appropriate, an electric door latch is also activated to grant access to the elder's dwelling for whoever the custodian calls to render assistance.

Some of the major issues found with these systems are:

1. Privacy is *very* important. The early systems nearly failed to gain acceptance because the elders were afraid that the custodian could listen into their private

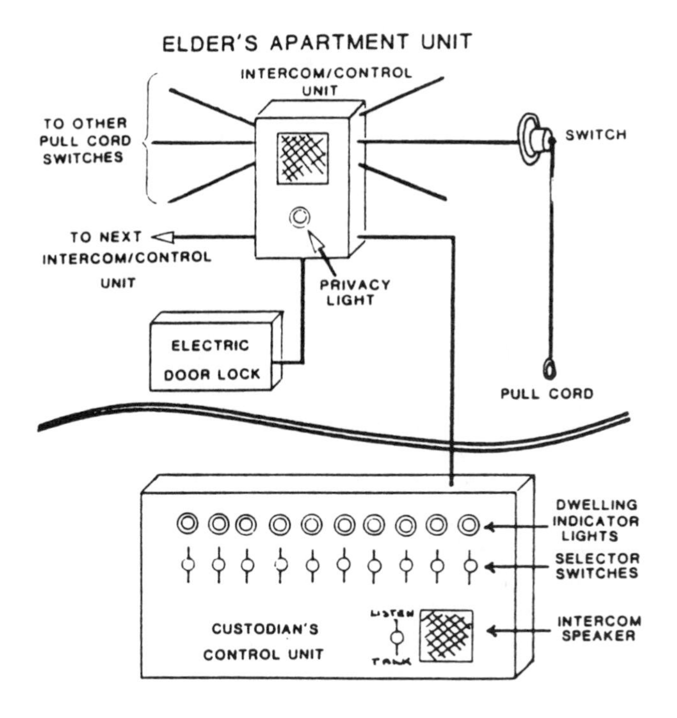

Figure 1. Typical English help call system.

conversations. This problem was solved by putting a red light on the intercom speaker which is lit when they can be overheard. TV cameras that allow visual monitoring by other people are usually unacceptable. Even under-mat switches that indicate movement within the apartment on the custodian's console have met with very mixed reception.

2. The calling mechanism must always be readily available. It would seem logical to equip the elders with a transmitting device so that they can always have the means to summon help immediately at hand. Experience has shown that these devices are not worn regularly over the long term and are often left in an inappropriate place such as a drawer. In some cases this may be due to forgetfulness; in other cases it may be denial. The pull cords are always there (they come with the building) and have proved to be effective in long term use.
3. Access control is important. In an emergency an elder may not be able to open the door and let the helpers in. In some installations the custodian has a master key to the elders' apartments, in others they do not, due to privacy concerns. Even in those cases where the custodian has a key, the custodian may have to handle several emergencies at the same time. Activating an entry way electric door latch whenever an alarm is sounded has proved to be an effective compromise between privacy and the need to grant access to help providers.
4. It must be simple to activate a call for help. Telephones are definitely too complex in times of stress and even pushbuttons may prove too difficult for arthritic hands.

Pull-cords are effective although they can be activated by pets (especially cats). Pets are of real emotional benefit to the elderly, but they do pose some additional problems to help-call systems designers.

5. The two way voice link is important for the custodian to be able to determine the type of help required. It is also important for eliminating false alarms which are quite frequent (due to accidental pulling of the pull-cords.) A simplex voice link is used with a talk/listen switch controlled by the custodian so that high sensitivity can be used in listening to the elder combined with a high volume when talking. This is important when the elder may not be able to speak very clearly in emergency (or may be in another room from the intercom) and may have impaired hearing.
6. The biggest weakness of these systems is that the elder may collapse without signalling an alarm. It is possible for an elder to lie unconscious for many hours before anyone realizes that anything is wrong. It is the responsibility of the custodian to monitor for this happening but this monitoring is limited by the elder's need for privacy and is often ineffective.

There are a significant number of infirm elderly who do not live in these government-provided apartment blocks for the elderly. A system that is provided for these people has pull-cords and an electric lock, but no intercom connection. Instead a flashing red light and alarm bell are used outside the dwelling to summon help from neighbors. This system has several limitations:

1. The response time may be long before anyone recognizes the alarm (in cities people tend to ignore alarms as "none of their business").
2. There is no trained person to respond to an emergency who has access to an infrastructure of help providers.
3. The elderly are reticent about using the system except in case of severe medical emergency for fear of "bothering the neighbors." This does not seem to be the case when a paid professional handles the calls.

Nonetheless, this system is certainly better than having no alarm system at all and can enable the infirm elderly to continue living in their own homes.

THE AMERICAN SCENE

In the United States, most infirm elderly do not live in government-provided housing. Many are institutionalized in nursing homes. Those fortunate enough to be relatively wealthy can live in somewhat more luxurious versions of the apartment buildings used in England. Some elderly live with their adult children often in problematic circumstances. Most live alone, many in inner cities, where they are supported by family, home-making services, meals-on-wheels, and visiting nurses. Monitoring is usually left to their adult children and fellow elderly.

The most widely used help-call system is that developed by Lifeline Systems, Inc.[1] This system provides its clients with a transmitting device that is worn on a cord around their neck or clipped to their belt. When the client needs assistance, he or she presses a button on this unit and a call for help is transmitted to an autodialer unit attached to the client's phone. This autodialer unit dials a minicomputer, which is typically located in a local hospital, and identifies the caller. The minicomputer then prints out the name and address of the client needing help.

Typically, the Lifeline system is manned by volunteers in the hospital or by hospital personnel as a secondary duty. These people will call a designated person, who is typically a neighbor, to check on the client and to summon help. The Lifeline subscriber unit will also automatically summon help if the subscriber has not used their telephone within a prescribed period of time. The Lifeline system is believed to have some 35,000 subscriber units in the field and over 100 hospitals offering the service to infirm elders who are discharged from the hospital. The Lifeline system does not have two way voice communication, dwelling access control, or paid professionals with a data base of help providers to handle the calls. It is, however, providing a needed link between the infirm elderly who live alone and help providers in their community.

Some security companies who provide fire and burglar alarms are now extending their systems to provide medical emergency calling over the same dedicated or call-up phone lines as are used for the burglar and fire alarm signalling. These organizations do have professional dispatchers on duty 24 hours a day to handle calls. The systems do not have two way voice or access control and it is an open question as to whether one can mix property protection (which is the primary function of these organizations) with helping people.

Those apartment buildings that are being built for the elderly using government subsidies usually have apartments equipped with emergency pull cords or push buttons. These systems sound an alarm outside the dwelling. They usually have no access control. There is no custodian to render assistance (or funds to pay for one).

We find that the help-call systems available to most elderly in America are functionally inferior to those in use in the elderly apartment complexes in England. There appear to be two reasons for this. First, despite the fact that the US Government spends more on the elderly than on defense,[2,3] the British Government plays a far more active role in caring for its elderly by extensive funding of elderly housing and by requiring that this housing meets their special needs. Second, the demographics are very different. England is a small island with concentrated population centers and a population which accept a high degree of government intervention in their lives. Despite obvious concentrations in the "Sun Belt," most American elderly live in their own homes scattered throughout our cities and towns.

One technological advantage that the American elderly do have is one of the best telephone services in the world. The service is reliable, low cost, and there is a telephone in just about every dwelling unit in the USA. This service is extensively used by the friends, relatives, and organized groups to monitor the well-being of the elderly. Unfortunately the telephone alone fails to meet a number of the major requirements for help call service such as ease of use, access control, and the guaranteed availability of someone to receive the call for help. The next section describes a system that is being built around the use of the telephone to provide a service that will be functionally equivalent or even superior to that available to the elderly residents of the English apartment buildings.

NEW DEVELOPMENTS

Help Call Systems Corporation (HCS) is developing a new help-call system that embodies the best of both the English and American experience. This system is shown in Figure 2. The HCS system functionally duplicates the equipment used in the English apartments with pull cords and a voice intercom. In this new system, these units are connected to a Microcomputer Control Unit which senses when help is needed and dials an HCS help-call center.

An HCS help-call center is intended to serve a city-wide area. It has dispatchers who man the center on a 24 hour a day basis and whose sole function is to arrange for help to be provided to clients of the help-call service. The dispatchers sit at computer terminals through which they have access to a data base of help providers which is customized for each subscriber. When a call comes in from a dwelling unit, a dispatcher

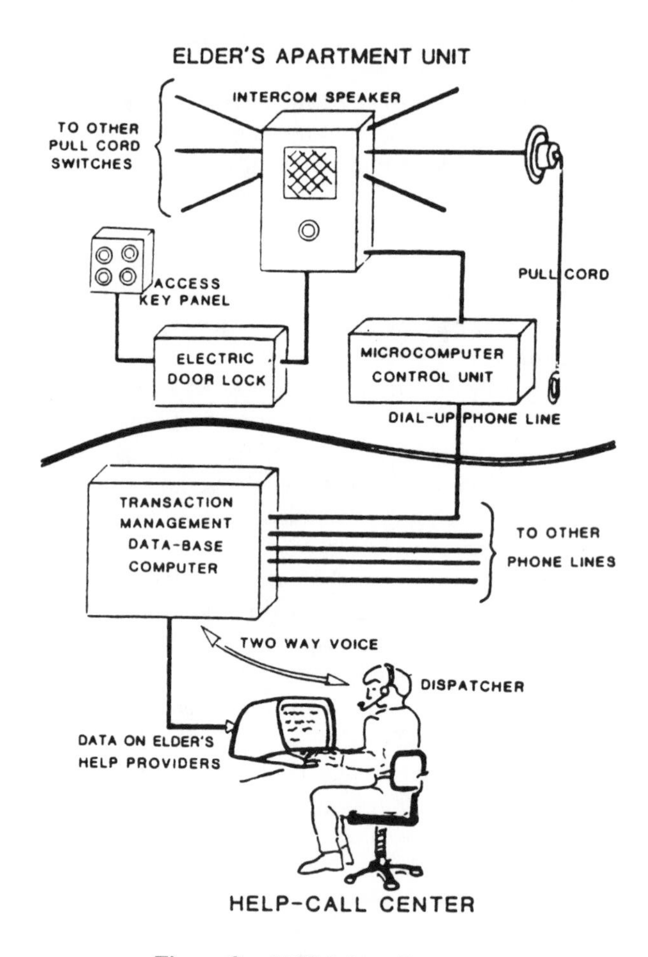

Figure 2. HCS help-call system.

is presented with information on the client such as name, address, pertinent medical history, and a list of help providers. The dispatcher is placed in voice contact with the client and determines what assistance is needed. The dispatcher then calls a help provider and activates the access lock so that the help provider may gain access to the client's dwelling. Finally, after the help provider has arrived, the dispatcher closes out the transaction by logging the disposition of the call (necessary for legal protection).

The access lock in this system has a keypad which is mounted near the access door to the dwelling. When an alarm is triggered by a client pulling on a cord, a unique code is generated and relayed electronically to the dispatcher. This code is given by the dispatcher to the help provider who enters it on the keypad when they arrive at the client's home. The code activates the lock and allows the help provider to gain access to the elder's dwelling. The code is unique and is only good for a limited time, such as one hour after the emergency. When the access code is entered on the keypad, a message is automatically relayed to the help-call center to inform the dispatcher that help has arrived for the client.

Access control is very important in the United States where the legal consequences of breaking into someone's home (even with good intentions) can be severe. The designers of this new system considered it unwise to simply unlock the dwelling in case of emergency as many elderly live in inner cities which tend to be high crime areas and they could be robbed before help arrived. With the code controlled access clients are assured that access will only be granted in case of emergency to those who are coming to render assistance. Also they are assured that the code will not be used for subsequent unauthorized access as it is only valid for the duration of the emergency.

One of the very important features of this system is the transaction-management data-base computer at the help-call center. This computer system manages all the telephone calls both from clients and to help-providers including answering, dialing, and logging and integrates these with the data base functions. The function of this system is to enable a few dispatchers to handle many clients in parallel, which is essential to the economical operation of a help-call center.

The time needed for a dispatcher to talk to a client and to a help provider is only a few minutes at most. All the rest of the work that would ordinarily be handled by the dispatcher such as looking up data on clients, dialing the phone numbers of helpers, and keeping track of whether help has arrived is handled by the computer. This includes the notification of the dispatcher of overdue events. When a dispatcher requests a help-provider to come to the assistance of a client, he enters an expected time for the help provider to arrive. If the help provider has not arrived within that time (i.e., not activated the access lock keypad), then the dispatcher is warned and can take appropriate action. There will be many telephone lines into a help-call center and a dispatcher may have a number of help-providers and even clients on-hold simultaneously. The system will warn the dispatcher if any of these have been on-hold too long and can automatically prioritize the dispatcher's actions in time of heavy activity.

Reliability is very important in these systems. The HCS system is designed with the Microprocessor Control Unit running off a trickle charged battery so that it can operate for at least 12 hours in the event of power failure. It is also heavily protected against power line surges and transients caused by lightning strikes on the telephone lines. The computer system at the help-call center uses a similar Microprocessor Control Unit to control each incoming and outgoing telephone line. These units are connected to a hot-standby pair of processors which are powered from battery-based uninterruptable power supply that is backed up by a manually started engine generator set. It is anticipated that the heaviest stress on the system will come with a city wide power failure accompanied by a lightning storm.

The use of the dial-up telephone line represents a possible weakness in the system but at present is the only financially viable communications medium available for use with widely scattered dwellings. Reliability of service is high, although the software and hardware used for the telephone interface has to take into account a significant probability of calls not being completed. The HCS system uses DTMF tone signalling mixed directly with voice for switching between talk and listen and also for keep alive messages so that the computers at both ends of the telephone link know that the other is still there. The same DTMF tones are used for handshaking by the computers when the dwelling unit dials into the help-call center and identifies itself. The use of DTMF tones makes for a low cost communications interface that can operate over a wide range of qualities of telephone lines.

The most economical way of achieving the complex logical functions required of the dwelling control unit was to use a microcomputer. This immediately opened up all sorts of possibilities to enhance the functionality and reliability of the system. The microcomputer is able to check on the correct functioning of much of the dwelling system such as mains power out and wiring malfunctions and to dial up the help-call center and automatically report these. It will be possible to tie in fire and burglar alarms as well as other more advanced sensing devices, some of which are described in the next section.

RESEARCH

One of the major shortcomings of all the systems, including the one under development by HCS, is that they depend on the client being able to activate a switch. If an elderly person suddenly collapses unconscious in the middle of the floor there is no way that any of these systems can help them. Monitoring for this condition by other people is undesirable because of the expense and especially because of privacy issues. Our research program at WPI is directed at developing systems that can automatically detect when a person needs help and call the help-call center.

There are three approaches to monitoring the well-being of the elderly and automatically calling for help:

1. Non-contact monitoring. In this approach, the motion of the elderly person about their home is tracked by ultrasonic, infrared, and pressure sensors. The person's well-being is then deduced from this motion, or lack of it.
2. Contact monitoring. In this approach, the elder wears devices in contact with his skin that are able to measure electrocardiological signals, pulse rates, respiration,

and the like. These signals are interpreted by a microcomputer which deduces the person's well-being and calls for help if needed.

3. Invasive monitoring. This is similar to item 2 above, except that the monitoring device is implanted in the client. A good example of this class of device is the newer heart pacemakers which are able to output cardiological data by means of a through-the-skin transducer.

All three cases require the use of real-time signal processing of the data from the sensors and real-time decision making to interpret the processed signals to determine whether the elder needs help. In addition contact and invasive monitoring require a transmitter to send data between the elder and the dwelling microcomputer. It is anticipated that the real-time signal processing will be done in the control unit carried by the elder and the signal interpretation will be done by the dwelling computer, which may perform interpretation based on all three modes of sensing.

Initially, we are focussing our attention on non-contact monitoring as this enables us to investigate the issues of real-time decision making without the complexities of portable monitoring devices or experiments with human subjects. We are currently investigating a number of potential sensors for monitoring including:

1. Door opening switches.
2. Under-mat and under-mattress pressure sensors that measure the weight applied as opposed to being simply switch closures.
3. Ultrasonic motion detectors that are able to measure range, motion, and to give an indication of the size of the object.
4. Infrared detectors that measure heat given off by people.

We plan to use the information from these sensors to detect whether the elder is in the dwelling and if so, where they are. Lack of motion of the client for a time will then be used as a primary indicator that help is needed. We also plan to incorporate the ability for the computer to speak to the client and to understand limited responses.

Currently, we are building an experimental people detection system using the ultrasonic rangefinding element from a Polaroid camera. This element is inherently low cost, although it does need a substantial amount of electronic circuitry which would have to be implemented in VLSI technology to keep the cost within acceptable bounds. In its simplest form, the electronics simply performs rangefinding using a burst of 50 KHz sound waves. We are now modifying this to allow the computer to examine the envelope of the returned burst which we hope to use to differentiate between moving people, pets, and stationary objects.

We are planning to integrate the data from the various sensors using a blackboard based artificial intelligence methodology.[4,5] This methodology formulates hypotheses based on sensed data and uses new data to improve or eliminate existing hypotheses. In this case the hypotheses will be about people and their movements. This method was selected as it is capable of being structured to run in real time, which many AI techniques are not.[6]

It is anticipated that in order to achieve a high probability of detecting an emergency, we will have to accept a high false alarm rate. For this reason we are planning to use voice response to ask ''Do you need help?'' and to listen to the response for ''Yes'', ''No'', ''Help'' and other simple phrases that can be understood by a speaker independent system. Ultimately this technology may evolve to where the computer interacts strongly with the client, greeting them when they get out of bed in the morning and warning when the range is left on (a common problem for many elderly).

The ultimate goal of all this research is to provide the technological assists needed for the elderly to live autonomous lives in their own homes. People are happier and live longer when they are surrounded by friends and neighbors and not institutionalized. The growing trend toward home care for the infirm reflects this.[7] These assists must be provided at an economical cost, however, if they are to be effective in practice which is why we are also performing research into developing new VLSI computer architectures that offer the potential of performing real-time decision making at low cost.

CONCLUSION

The help-call systems in use in government provided apartment blocks in England have been described and found to be functionally superior to the help-call systems available to most American elderly. A new help-call system under development by Help Call Systems Corporation has been described and shown to offer a higher level of functionality for the American elderly than is available in the English apartment systems. Finally, current research into the use of advanced sensing and computer techniques has been described, through which it is hoped to develop techniques that will allow the elderly to lead autonomous lives without fear of collapsing and lying for many hours without help.

REFERENCES

1. L.D. Shapiro, ''Lifeline Systems, Inc.,'' Form S1, Registration Statement to the SEC, Washington, DC, 27 May 1983.
2. ''Driving down the costs of an aging America,'' *Business Week*, 26 March 1984, pp. 58–60.
3. ''The defense buildup doesn't have to happen all at once,'' *Business Week*, 26 March 1984, p. 64.
4. L.D. Erman, F. Hayes-Roth, V.R. Lesser, and D.R. Reddy, ''The Hearsay-II Speech Understanding System: Integrating Knowledge to Resolve Uncertainty,'' *Computing Surveys*, vol. 12, no. 2, June 1980, pp. 312–353.
5. D.D. Corkill and V.R. Lesser, ''A Goal-Directed Hearsay-II Architecture: Unifying Data-Directed and Goal-Directed Control,'' Technical Report 80-15, Department of Computer and Information Science, University of Massachusetts, Amherst, Mass., June 1981.

6. P.E. Green, ''Resource Control in a Real-Time Target Tracking Process,'' *Proceedings of the Fifteenth Asilomar Conference on Circuits, Systems, and Computers,* Pacific Grove, Ca., November 1981.
7. K. Dawson and A. Feinberg, ''Hospitals in the Home,'' *Venture*, August 1984, pp. 38–43.

Index